SWITCH-LEVEL TIMING SIMULATION OF MOS VLSI CIRCUITS

THE KLUWER INTERNATIONAL SERIES IN ENGINEERING AND COMPUTER SCIENCE

VLSI, COMPUTER ARCHITECTURE AND DIGITAL SIGNAL PROCESSING

Consulting Editor

Jonathan Allen

Other books in the series:

Logic Minimization Algorithms for VLSI Synthesis. R.K. Brayton, G.D. Hachtel, C.T. McMullen, and A.L. Sangiovanni-Vincentelli. ISBN 0-89838-164-9.
Adaptive Filters: Structures, Algorithms, and Applications. M.L. Honig and D.G. Messerschmitt. ISBN 0-89838-163-0.
Introduction to VLSI Silicon Devices: Physics, Technology and Characterization. B. El-Kareh and R.J. Bombard. ISBN 0-89838-210-6.
Latchup in CMOS Technology: The Problem and Its Cure. R.R. Troutman. ISBN 0-89838-215-7.
Digital CMOS Circuit Design. M. Annaratone. ISBN 0-89838-224-6.
The Bounding Approach to VLSI Circuit Simulation. C.A. Zukowski. ISBN 0-89838-176-2.
Multi-Level Simulation for VLSI Design. D.D. Hill and D.R. Coelho. ISBN 0-89838-184-3.
Relaxation Techniques for the Simulation of VLSI Circuits. J. White and A. Sangiovanni-Vincentelli. ISBN 0-89838-186-X.
VLSI CAD Tools and Applications. W. Fichtner and M. Morf, editors. ISBN 0-89838-193-2.
A VLSI Architecture for Concurrent Data Structures. W.J. Dally. ISBN 0-89838-235-1.
Yield Simulation for Integrated Circuits. D.M.H. Walker. ISBN 0-89838-244-0.
VLSI Specification, Verification and Synthesis. G. Birtwistle and P.A. Subrahmanyam. ISBN 0-89838-246-7.
Fundamentals of Computer-Aided Circuit Simulation. W.J. McCalla. ISBN 0-89838-248-3.
Serial Data Computation. S.G. Smith and P.B. Denyer. ISBN 0-89838-253-X.
Phonological Parsing in Speech Recognition. K.W. Church. ISBN 0-89838-250-5.
Simulated Annealing for VLSI Design. D.F. Wong, H.W. Leong, and C.L. Liu. ISBN 0-89838-256-4.
Polycrystalline Silicon for Integrated Circuit Applications. T. Kamins. ISBN 0-89838-259-9.
FET Modeling for Circuit Simulation. D. Divekar. ISBN 0-89838-264-5.
VLSI Placement and Global Routing Using Simulated Annealing. C. Sechen. ISBN 0-89838-281-5.
Adaptive Filters and Equalisers. B. Mulgrew, C.F.N. Cowan. ISBN 0-89838-285-8.
Computer-Aided Design and VLSI Device Development, Second Edition. K.M. Cham, S-Y. Oh, J.L. Moll, K. Lee, P. Vande Voorde, D. Chin. ISBN: 0-89838-277-7.
Automatic Speech Recognition. K-F. Lee. ISBN 0-89838-296-3.
Speech Time-Frequency Representations. M.D. Riley. ISBN: 0-89838-298-X.
A Systolic Array Optimizing Compiler. M.S. Lam. ISBN: 0-89838-300-5.
Algorithms and Techniques for VLSI Layout Synthesis. D. Hill, D. Shugard, J. Fishburn, K. Keutzer. ISBN: 0-89838-301-3.

SWITCH-LEVEL TIMING SIMULATION OF MOS VLSI CIRCUITS

by

Vasant B. Rao
David V. Overhauser
Timothy N. Trick
Ibrahim N. Hajj

University of Illinois

Kluwer Academic Publishers
Boston/Dordrecht/London

Distributors for North America:
Kluwer Academic Publishers
101 Philip Drive
Assinippi Park
Norwell, Massachusetts 02061 USA

Distributors for the UK and Ireland:
Kluwer Academic Publishers
Falcon House, Queen Square
Lancaster LA1 1RN, UNITED KINGDOM

Distributors for all other countries:
Kluwer Academic Publishers Group
Distribution Centre
Post Office Box 322
3300 AH Dordrecht, THE NETHERLANDS

Library of Congress Cataloging-in-Publication Data

Switch-level timing simulation of MOS VLSI circuits / by Vasant B. Rao
... [et al.].
p. cm. -- (The Kluwer international series in engineering and
computer science ; 66. VLSI, computer architecture and digital
signal processing)
Includes index.

1. Integrated circuits--Very large scale integration--Computer
simulation. I. Rao, Vasant B. II. Series : Kluwer international
series in engineering and computer science ; SECS 66. III. Series :
Kluwer international series in engineering and computer science.
VLSI, computer architecture and digital signal processing.
TK7874.S87 1989
621.381 '73 '0724--dc19 88-30591
CIP

ISBN-13: 978-1-4612-8963-0 e-ISBN-13: 978-1-4613-1709-8
DOI: 10.1007/978-1-4613-1709-8

Softcover reprint of the hardcover 1st edition 1989

CONTENTS

PREFACE

Only two decades ago most electronic circuits were designed with a slide-rule, and the designs were verified using breadboard techniques. Simulation tools were a research curiosity and in general were mistrusted by most designers and test engineers. In those days the programs were not user friendly, models were inadequate, and the algorithms were not very robust. The demand for simulation tools has been driven by the increasing complexity of integrated circuits and systems, and it has been aided by the rapid decrease in the cost of computing that has occurred over the past several decades. Today a wide range of tools exist for analysis, design, and verification, and expert systems and synthesis tools are rapidly emerging. In this book only one aspect of the analysis and design process is examined, but it is a very important aspect that has received much attention over the years. It is the problem of accurate circuit and timing simulation.

In the early days of integrated circuits, logic simulators and circuit simulators were the primary tools of interest. Logic simulators performed Boolean operations and could handle relatively large digital circuits, but yielded crude timing information at best. Circuit simulation came of age with the wide distribution of the general purpose circuit simulator SPICE (Simulation Program with Integrated Circuit Emphasis) in the 1970's. This program was developed by faculty and students at the University of California in Berkeley, California. It has become the dominant circuit simulation program, because it was extensively tested in industry and distributed for a nominal handling charge.

However, as soon as circuit simulation began to become an accepted tool, it was realized that it was too slow to handle the complexity of modern integrated circuits. In 1975, Bell Laboratories announced MOTIS (MOs circuit TIming Simulator). This program used table look-up techniques for transistor models, and, because the program was restricted to MOS circuit analysis, relaxation methods could be successfully used. In a restricted design environment this program could generate accurate voltage and current waveforms an order of magnitude faster than a general purpose simulator.

Logic simulation was not marking time in this era. As MOS circuit design progressed, it became clear that logic simulators could not adequately describe the digital behavior of pass transistors. As a result, switch-level simulators, such as MOSSIM, were developed in which the MOS transistor is represented as a switch whose state is controlled by the gate of the transistor. The switch can be either on, off, or in the undefined state. Since these programs only perform Boolean operations, they typically are three orders of magnitude faster than general purpose circuit simulators. Timing information can be generated by estimating the grounded capacitance at each node and replacing transistors by an equivalent resistance. RC network reduction techniques can be used to estimate the time constant of the connected paths. Although these techniques are very fast, typically two to three orders of magnitude faster than a general purpose circuit simulator, they can give significant timing errors.

In this book IDSIM, a switch-level timing simulator for MOS circuits, is presented which yiëlds superior timing and waveform information, and yet is still more than two orders of magnitude faster than a general purpose circuit simulator. It is like a switch-level simulator in that it uses network reduction techniques to reduce dc connected paths to a single equivalent transistor to ground and a single equivalent transistor to the supply voltage. However, rather than solve the nonlinear differential equation for this equivalent circuit, as is done in circuit simulation in which the time step is controlled by the local truncation error, this simulator generates an approximation to the charging current at the output node, and determines the time required for the output voltage to cross a specified threshold. Two voltage thresholds are specified so that accurate slope information for the waveform is generated. A novel dynamic windowing technique is presented to handle feedback loops.

After the Introduction, Chapter 2 presents an overview of circuit, logic, and timing simulation. Chapter 3 presents algorithms for partitioning and ordering MOS circuits for switch-level simulation. Section 3.3 explains how to identify pass transistors. Most switch-level simulators only partition MOS circuits into channel-connected blocks. However, to obtain improved timing estimates it is necessary to identify pass transistors from the pullups and pulldown transistors. Chapter 4 presents the algorithms for fast and accurate timing analysis. The dynamic windowing algorithm for feedback loops is the subject of Chapter 5. Finally, some performance results are presented in Chapter 6 which demonstrate the speed and accuracy of the simulator IDSIM.

This book is based on the Ph. D. dissertations of Vasant Rao and David Overhauser. Our special thanks to Lilian Beck for editing the manuscript.

V. B. Rao
D. V. Overhauser
T. N. Trick
I. N. Hajj

SWITCH-LEVEL TIMING SIMULATION OF MOS VLSI CIRCUITS

CHAPTER 1

INTRODUCTION

The design of an electronic circuit, traditionally, started with the designer who, with a mental picture, translated his or her ideas into the form of a circuit schematic. This step relied heavily on the human designer's intuition, past experience, and knowledge to make reasonable approximations. This was followed by the "breadboarding" phase in which an actual prototype of the circuit was constructed from discrete components interconnected by external wires and was tested. The performance of the circuit, if not found satisfactory, was then improved by adjusting the circuit element values in a somewhat trial-and-error fashion.

The advent of integrated circuits, however, has greatly changed the picture. There are several steps involved in the design of a *very large-scale integrated* (VLSI) circuit, which may consist of several hundreds of thousands of components, mainly transistors. The circuit designer first obtains a very high-level functional description of the circuit based on specifications provided by the user. The synthesis, often called the top-down process, translates this high-level description into various levels including the register level, the transistor level (or electrical level) etc. and terminates at the physical mask-level, i.e., the actual layout of the patterns of metal, semiconductor, and insulating material by which the components and the interconnections are achieved. This is followed by the design verification, or the bottom-up process, wherein a software tool called an extractor is first used to obtain a circuit level (or transistor level) description from the physical layout. The breadboarding phase is replaced by using a simulation tool to predict the performance of the circuit which is then compared with the user's specifications; thus, completing the so-called design loop. If the performance is not satisfactory, certain changes are made and the whole process is then repeated. The total time spent in the design loop is usually referred to as the *turn-around time*.

The main objective of the VLSI circuit designer is to obtain designs with as low a turn-around time as possible. Computer-aided design (CAD) tools have become virtually indispensable at various steps in the design process to perform tasks which would otherwise take a very long time if they were done by human beings. Using silicon compilers can speed up the top-down synthesis process considerably since they produce the mask level description, directly, from the functional description without any human intervention. Certain software tools known as *design rule checkers* (DRC) and *electrical rule checkers* (ERC) are also used. These perform the rather mundane tasks of checking to see if the layout satisfies all the design rules of the technology and whether there are any topological faults from the electrical point of view such as a floating node, or a short between power and ground. There is, however, a bottleneck in speeding up the bottom-up design verification process. This bottleneck is in the simulation of the electrical behavior of the circuit due to the unavailability of a simulation tool that is capable of accurately predicting the performance of an entire VLSI circuit at a reasonable cost. The accuracy of a simulator is important, since otherwise the integrated circuit which is fabricated and tested might turn out to perform rather unsatisfactorily. For large circuits (typically of the kind in today's VLSI technology), the speed of simulation is equally important so that the entire circuit can be simulated in a reasonably small amount of computation time. However, speed and accuracy of a simulator are often conflicting requirements among existing simulation tools.

In this book we will be primarily concerned with providing a fast and accurate simulation tool to a VLSI circuit designer that gives adequate information on the performance of the circuit with a reasonable expenditure of computation time even for very large circuits. Most of the existing simulators for integrated circuits can be classified into two distinct categories, namely, *analog simulators* and *digital simulators*. Analog simulators treat an electronic circuit as a continuous dynamical system with electrical signals such as voltages and currents. Digital simulators, on the other hand, view the circuit as a digital network with signals occupying discrete states such as low (0) and high (1). For small circuit blocks where analog voltage levels are critical in evaluating circuit performance, or where strong coupling exists, analog circuit simulators such as SPICE2 [1] and ASTAP [2] can be used to predict the performance of the circuit very accurately. As the size of the circuit (number of components) increases, however, using these simulators is no longer cost-effective. Several decomposition techniques have been used to speed up their performance and have resulted in a new generation of analog simulators [3-15] which are, however, cost-effective for circuits limited to at most ten-thousand devices.

The existing digital simulators [13-27] can be further divided into Boolean gate-level [13-18] and switch-level [19-27] simulators. In the Boolean gate model a circuit consists of a set of logic gates connected by unidirectional memoryless wires. The logic gates compute Boolean functions of their input signals and transmit these values along the wires to the inputs of other gates. Each gate input has a unique signal source. Information is only stored in the feedback paths of sequential circuits. The Boolean gate model, however, cannot describe some of the newer technologies currently used in VLSI circuit design, especially circuits with Metal-Oxide-Semiconductor (MOS) transistors. The MOS transistor can be treated as a voltage-controlled switch with three terminals : drain, gate, and source. The signal at the gate terminal controls the connection between drain and source terminals. Therefore, some MOS pass transistor networks can implement combinatorial logic in ways that resemble relay contact networks more closely than conventional logic gate networks. Dynamic memories using MOS devices can store information without feedback paths by exploiting the capacitance of the wires (interconnect region) and the gates of the transistors attached to them. A variety of bus structures can provide multidirectional, multipoint communication. Thus,

MOS circuits consist of bidirectional switching elements connected by bidirectional wires with memory due to the interconnect and device capacitances. Hence, they cannot be modeled accurately by Boolean gate-level simulators.

A new class of digital simulators has recently emerged specifically for simulating MOS VLSI circuits. These switch-level simulators [19-27] model an MOS circuit as a set of nodes connected by transistor switches. Each node occupies a discrete number of states **0**, **1**, or **X** for the intermediate or unknown state and each switch is either *open*, *closed*, or in an *intermediate* or *unknown* state. These simulators can handle a variety of MOS configurations such as logic gates, pass transistors, busses, static and dynamic memory. Digital simulators, in general, operate at sufficient speeds to test entire VLSI systems, since the circuit behavior is modeled at a logical rather than a detailed electrical level. However, these simulators do not model the dynamics of the circuits properly and are often useful only in predicting steady-state responses of the signals. Analog simulators, on the other hand, predict both steady-state and transient responses fairly accurately, if the device models used are accurate, but are cost effective only for circuits, with less than a few thousand components, which are considered small in the present day VLSI technology.

The algorithms presented in this book have led to the development of IDSIM; a switch-level timing simulator for MOS VLSI circuits which attempts to bridge the gap between analog and digital simulators. IDSIM performs simulations at a switch level; hence it runs at speeds close to that of digital simulators. Furthermore, it uses a delay operator to delay signal transitions accurately, and therefore, provides the timing accuracy comparable to that of analog simulators. The input to IDSIM is a transistor-level circuit description in a SPICE2 input format. The program begins by partitioning the entire MOS network into several channel-connected functional blocks. The partitioning is an automatic process that is completely transparent to the user. The partitioned blocks are then ordered for processing so that, whenever possible, a block is scheduled for processing only after all its inputs have been previously processed. Since this is not possible for blocks forming feedback loops, a novel dynamic windowing scheme is used to schedule such blocks. The blocks are then processed at a switch level and signal transitions are delayed suitably by a delay operator.

In Chapter 2 we begin with an overview on existing digital simulation techniques. In Chapter 3, the algorithms for partitioning the input network into various blocks and the ordering of these blocks for processing are discussed. The partitioning and ordering steps are common to most switch-level simulators; hence, these are discussed in a general setting. In this chapter we also present algorithms to identify pass transistors in the MOS network, in case this is required for simulation. Chapter 4 deals with the specific algorithms in IDSIM for the switch-level timing simulation of MOS circuits. Chapter 5 describes the details of the novel dynamic windowing technique used to process blocks within feedback loops. Finally, in Chapter 6, we discuss the implementation of IDSIM and present some results.

CHAPTER 2

OVERVIEW OF SIMULATION TECHNIQUES

2.1 ANALOG vs DIGITAL SIMULATION

Simulation plays a major role in the process of designing an integrated electronic circuit. By using a simulator, the circuit designer can evaluate the performance of the design before going into the expensive and time-consuming manufacturing process. There are two basic approaches to simulating an integrated electronic circuit. The first, and more traditional, approach is to treat the circuit as a continuous dynamical system and obtain a set of nonlinear algebraic-differential equations with electrical variables such as voltage, current, and charge to describe its behavior. The objective of an *analog simulator* is to solve this set of equations, numerically, and obtain the detailed waveforms at various nodes in the circuit.

For small circuit blocks where analog voltage levels are critical to determine circuit performance, or where strong coupling exists, circuit simulators such as SPICE2 [1] and ASTAP [2] can be used to provide accurate information on the behavior of the circuit. These simulators are often referred to as *standard circuit simulators*. These are general purpose simulators in that they can handle almost any type of circuit element such as resistors, capacitors, inductors (both self and mutual), voltage and current sources (independent and controlled), nonlinear devices (transistors, diodes, etc.), and transmission lines. They can also perform many types of analyses such as *dc* analysis, *ac* (or small-signal) analysis, noise analysis, and transient or time-domain analysis. Several attempts have been made to speed up the performance of standard circuit simulators. This has resulted in the development of a variety of analog simulators such as SLATE [3], MACRO [4], MOTIS [5], MOTIS-C [6], MOTIS-II [7], PREMOS [8], RELAX [10-12], SPLICE [13], DIANA [14], and SAMSON [15].

An alternate approach is to view the circuit as a digital system in which the signals occupy discrete states. Since the majority of VLSI circuits are primarily digital in nature, *digital simulators* are often successful in predicting steady-state responses in these circuits. Analog simulators are generally quite accurate in evaluating the performance of circuits, but are not fast enough to handle entire VLSI circuits. Digital simulators, on the other hand, are able to simulate very large circuits, but, unfortunately, are not accurate in modeling the dynamics in these circuits.

Digital simulators [13-28], or logic simulators as they are often called, form an important class of computerized tools for designing very large integrated circuits. These simulators provide a discrete "on/off" type analysis of the circuit under test. Signal values are described by a fairly small number of discrete levels rather than in a continuous range as is the case in an analog simulator. Through the use of very simple models for the devices and Boolean arithmetic to perform operations on the discrete signal values, digital simulators are often capable of economically analyzing circuits containing the equivalent of over 100k active devices. The dynamics of the circuit are, however, modeled by simply *delaying* the various signal-transitions between the discrete levels. In most cases a simple, user-defined rise and fall delay between the input and output of a logic-gate or transistor-group is used. Thus, digital simulators, at best, provide a fairly crude, first-order timing analysis of the circuit under consideration.

Digital simulators are useful and popular since most integrated circuits are primarily digital in nature. The usefulness of a digital simulator, however, depends greatly on the consistency and accuracy with which it can model the logic behavior of a full range of design techniques available to the designers of integrated circuits. Of course, no digital simulator can model all designs with complete accuracy, because it does not simulate the detailed analog behavior of the circuit. It should, nonetheless, provide as close a model as possible within a set of well-defined limitations. As a further requirement, a digital simulator for VLSI circuits must be efficient enough to simulate entire systems with reasonable speed. A digital simulator has, as its basis, an abstract *logical model* that describes both the structure and the behavior of a system in terms of a set of primitive elements, a set of interconnections, and a set of rules for operation. For a simulator to accurately and reliably simulate a system, the logical model must reflect its actual structure and operation. Digital simulators can be divided into two categories, namely, Boolean gate-level simulators [13-22] and switch-level simulators [23-28].

2.2 GATE-LEVEL SIMULATION

The Boolean logic gate model has formed the theoretical basis for logic design ever since the advent of electronic logic. In this model a circuit is composed of several logic gates connected by unidirectional, memoryless wires. The logic gates themselves are collections of transistors and/or other circuit elements which perform a logic function. A logic gate may be a simple inverter, NAND gate, or NOR gate, or a more complex functional unit such as a flip-flop or a register. The logic gates compute Boolean functions of their input signals and transmit these values along wires to the inputs of other gates to which it might be connected. Each gate input has a unique signal source. Information is stored only in feedback paths of sequential circuits. The Boolean gate model directly implements the well-known two-valued Boolean algebra, and consequently has a well-defined specification which can guide the simulator implementation.

The unilateral nature of logic gates is fundamental to the operation of gate-level simulators. For each binary vector at the input nodes of a logic gate, the binary value (i.e., **0**, or **1**) at the output is computed and propagated on to the inputs of other gates that might be connected to it. Due to the inertial elements such as node capacitances present in the circuit, however, a change in the state of the input to a gate would propagate to the output only after a certain time-delay. Simulators which do not account for this delay can analyze only combinatorial circuits. Thus, simulators which handle sequential circuits must estimate the propagational delay through a logic gate and they do so in several ways. Some simulators operate in the so-called *unit delay* mode, where all logic gates are assumed to have the same delay. Unit-delay simulators, however, can verify only the steady-state behavior or the logic functionality of the digital circuit. In order to provide some kind of timing information, some simulators allow *assignable delays* where the user can assign specific delays through any of the logic gates used in the simulation. Even in assignable delay simulators, the delay values may only be integer multiples of a fundamental time-quantum, usually referred to as the *minimum resolvable time* (MRT). For example, the MRT in a certain simulator may be 0.1 ns, in which case a gate delay of 10 units represents an effective delay of 1.0 ns.

The difference in propagational delays through different signal paths in a network of logic gates may sometimes cause undesirable situations, such as *static hazards* and *dynamic hazards*. Hazards [40-44] are situations where it is possible for spurious glitches or spikes to appear in an otherwise smooth analog waveform at the output of a logic gate. In a sequential circuit, the occurrence of a glitch could cause the circuit to malfunction. Therefore, the detection of hazards and race conditions are very important, and as a result, most digital simulators caution the user when they occur. The detection of hazards is possible by introducing a third state, usually denoted by X, to represent signal-transitions [40-44].

The Boolean gate model cannot represent many of the design techniques currently used in VLSI design. This is especially true in the case of MOS VLSI circuits, as explained in Chapter 1. Hence, most existing digital simulators extend the Boolean gate model in various ways to handle MOS circuits.

Many simulators extend the two-valued logic of Boolean algebra with a third value to represent an *unknown* or *undefined* logic level. This X state could indicate an uninitialized signal, a signal held between two logic thresholds, or a signal in a 0→1 or 1→0 transition. The X state is handled algebraically by extending the binary Boolean algebra to a ternary or three-valued De Morgan's algebra [18,40] which preserves many of the desirable mathematical properties of the Boolean gate model. The X state implemented this way is also useful in the detection of hazards and race conditions [40–44]. Alternatively, some simulators implement the X state by an enumeration technique in which the simulation is repeated with the nodes in the X state set to all possible combinations of 0's and 1's [22]. Nodes that remain in a unique binary state for all combinations are set to this state, while all others are set to X. To simulate tristate gates and logic busses, some simulators use a fourth state, called the *high impedance* state, and often denoted by H [16]. This H state is also used sometimes to model dynamic memory by allowing a node to retain its previous logic state if the outputs of all logic gates connected to the node are at the H level.

As far as simulation is concerned, most gate-level simulators belong to one of two general types. The first is based on the Huffman logic model [19], as shown in Figure 2.1. In this model, all the feedback paths in the network are initially broken resulting in a purely combinatorial network, which is then levelized in terms of signal dependence. The feedback is restored by inserting delay elements between the secondary outputs and secondary inputs of the combinatorial part of the

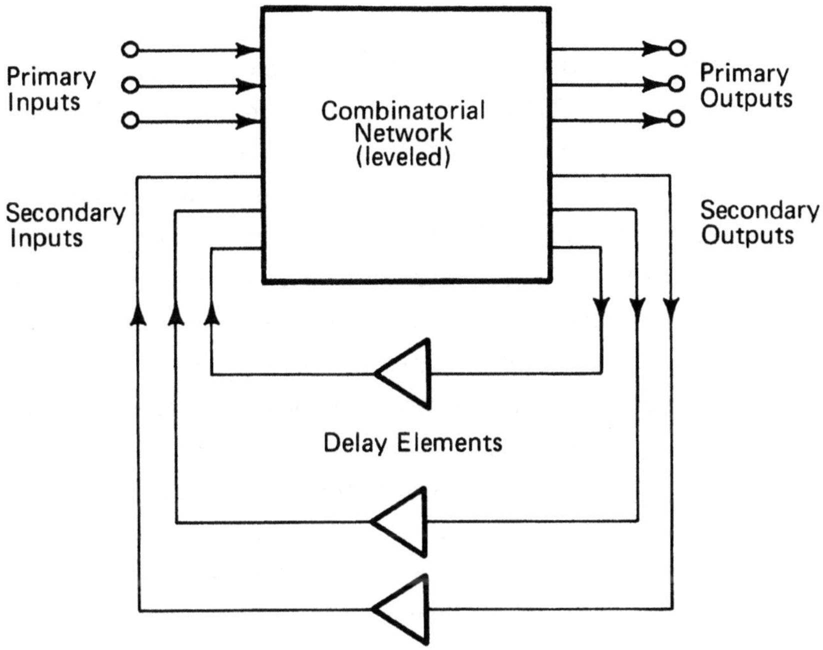

Figure 2.1 : The Huffman logic model for logic analysis

network. The analysis begins by applying the input excitations and following paths where the signal states change through the network to the outputs. The delays are applied to any secondary output change and the analysis of the combinatorial block begins once again. The process is repeated until the requested input sequence has been completed. This approach is used in

SALOGS [16], and is quite efficient for circuits in which relatively few delays are significant, or in other words, for nearly combinatorial circuits.

The second and more common approach is based on the use of a time-queue (TQ) [20,21] as shown in Figure 2.2. Each entry in the queue represents a discrete point in simulation time. Time moves ahead in fixed increments which correspond to consecutive entries in the TQ. Each entry in the queue contains a pointer to a list of *events* which are to occur at that instant of time. An event is usually defined as a change in the logical state of an output node of an *element*. The element, in this case, may be a voltage source or a logic gate. The new state may or may not be the same as the state already held by the output line. If the new state is different from the old one, then all elements whose input lines are connected to this output line, called *fanout* elements, must be processed to see if this change affects their outputs. If an element gets processed at say, time t_i, and the input event is found to cause an output event, then the output event is assumed to occur at time t_{i+k} where $k>0$ represents a *positive delay* through the logic gate. The fanouts of the output node then get scheduled for processing at time t_{i+k}. If the state of an output node remains unchanged, then the fanouts are not added to the time-queue. This approach is often referred to as a *selective trace* technique, or an *event-driven* scheme, or sometimes even as *dynamic leveling*. In the case of logic simulation, no penalty in accuracy or stability of analysis is incurred with the use of the selective trace method. One of the advantages of this scheme is that it allows different gates to have different delays and, moreover, the delay value

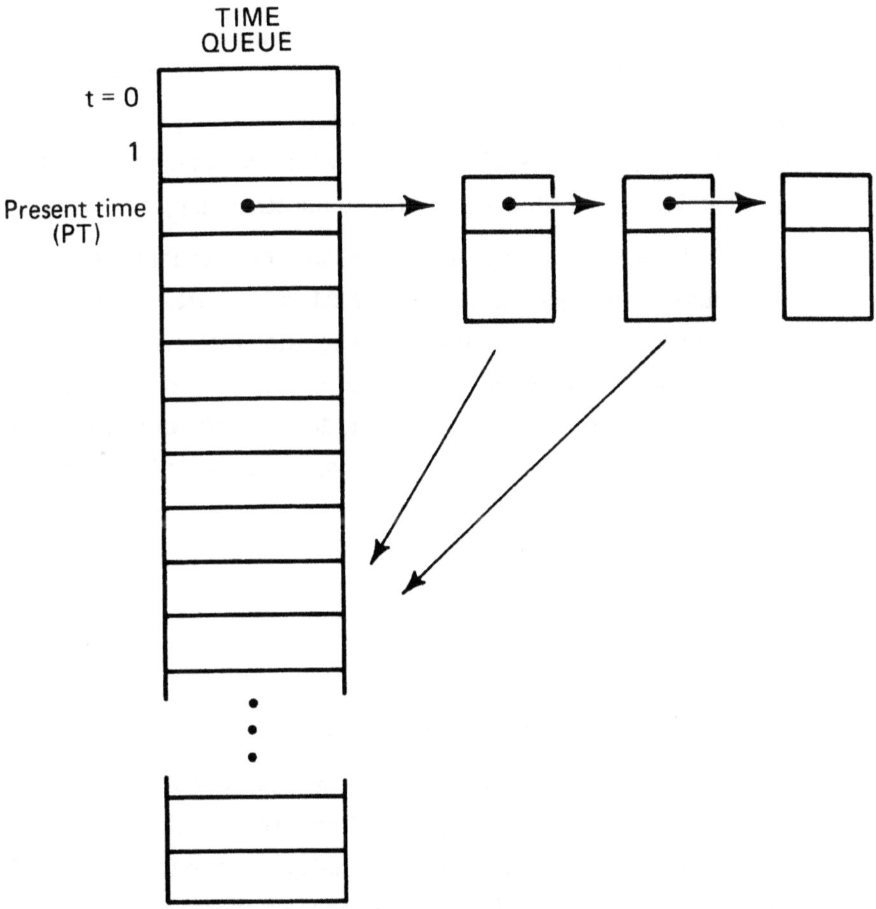

Figure 2.2 : The principle of the time-queue simulator

through a gate is also allowed to change as the simulation proceeds. This is especially good for MOS logic gates which have different delays for *rising* transitions and *falling* transitions at the output respectively. Furthermore, the presence of feedback among the logic gates does not complicate the simulation, since the delay through a feedback loop would schedule a gate for processing only in the future and never at the present time in the queue. Several logic simulators such as TEGAS [17], and SPLICE [13], use the TQ approach successfully.

Gate-level simulators, however, are not entirely suitable for the digital or logic simulation of MOS circuits. This is due to the fundamental mismatch between the Boolean gate model and the behavior of MOS logic circuits. Hence, the need for a different approach to the digital modeling and simulation of MOS circuits is apparent and is discussed in the following section.

2.3 SWITCH-LEVEL LOGIC SIMULATION

A class of digital simulators known as switch-level simulators has emerged recently as an alternative to the more conventional gate-level simulators, especially for the simulation of MOS circuits. One of the first switch-level simulators to be implemented is MOSSIM [23]. Since then a large number of switch-level simulators have been implemented. A list of some of these can be found in [64].

In contrast to gate-level modeling and simulation, switch-level techniques operate directly on the transistor circuit structure and capture many circuit properties that are hard, if not impossible, to model at the gate level, such as bidirectionality of signal flow, charge sharing effects, and transistor sizes. In contrast to circuit-level simulation, node voltages are represented by discrete logic levels and transistors by *bidirectional resistive switches* in switch-level modeling. Thus, an MOS circuit is modeled by a set of nodes connected by resistive switches. Node voltages are represented by a small number of discrete logic levels; in most cases they are: **0** (low voltage), **1** (high voltage), and X or U (undefined or unknown, i.e., any value between and including **0** and **1**). In some cases the high impedance state Z is also included [65].

The resistive switch model of a transistor is controlled by the voltage level at its gate terminal. An n-type transistor is conducting when its gate voltage is **1**, and a p-type transistor is conducting when its gate voltage is **0**. Transistors are allowed to have discrete *strength values* depending on the values of their conductances when fully on. This is done in an effort to model the behavior of ratioed logic. For example, a depletion load transistor used in n-channel MOS (NMOS) circuit design has its gate logic level set to **1** and its strength is weaker that that of an enhancement type transistor. Transistors in series are equivalent to a single transistor of strength equal to the weakest one (minimum conductance), while transistors in parallel are equivalent to a single transistor of strength equal to the strongest one (maximum conductance).

Input nodes, such as power, ground, clock drivers, and data nodes, are considered to have specified logic levels and are capable of influencing the states of the other nodes in the circuit, but not each other. All other nodes are considered storage nodes that have capacitances capable of storing charges. Storage nodes are assigned discrete sizes, depending on their relative capacitance values. Node sizes come into play when storage nodes become disconnected from input nodes. In this case charge sharing occurs, and it is assumed that the nodes of the highest size dominate the charge on nodes of smaller size connected to them.

In most, if not all, switch-level simulators the circuit is partitioned into channel-connected subcircuits. This partitioning could be done statically once at the outset, where every transistor is included; or dynamically at every iteration where only conducting transistors are considered. This dynamic partitioning usually produces smaller subcircuits, but at an additional overhead cost. Input voltages, such as supply voltage and ground, which have specified voltage levels, logically disconnect the subcircuits from each other. The interaction between subcircuits occurs only at transistor gate terminals. The simulation of the entire circuit follows an event scheduler similar in many ways to gate-level logic simulators, except that now the gates consist of channel-connected transistors.

Given the input states and initial conditions of a subcircuit, the resultant steady-state solution can be found in a number of ways. In the method described by Bryant [25], the node sizes and the transistor strengths are ordered in ascending order of strength such that the storage nodes are ordered first followed by the transistor strengths, with the highest strength reserved for the input node. The strength of an input signal source as seen by a storage node through a path of conducting transistors is determined by the minimum transistor strength in the path, while the strength of a signal from another storage node is determined by the size of that storage node. The ordering of the node sizes and transistor strengths is used to determine the logic level at a node. Some of the paths arriving at a node might be blocked [25] by other paths. A path is said to be *blocked* at a node if there is another path from that node to an input node which is stronger. Blocked paths are excluded from consideration when determining the logic level at the node. The new state at the storage node in question is then found by applying the *connection operator* defined by Hayes [65]. Path tracing graph algorithms are applied to determine the signal strengths arriving at the node [25,26,28]. More recently, matrix formulation and Gaussian elimination techniques have been employed [66,67]. In most cases iterative methods are used to arrive at the steady-state solution. Convergence can be speeded up by considering signals in steps according to their strengths [26,28].

When some transistors have their gate values at X, special techniques are employed to arrive at a correct solution. A straightforward approach would be to solve the subcircuit under consideration for all possible combinations of X put to 0 and 1. Those nodes that have their steady-state values unchanged under all combinations will have their states set to those values; otherwise, the state will be X. However, such an approach has an exponential complexity. As a result, several methods have been proposed to reduce the complexity, where only two iteration sequences are required to obtain the correct result [25,26,28]. In order to keep the computation simple and fast, circuit delays are not accurately modeled. A simple unit delay is often used. Arbitrary delays or delays with specified rise and fall times can also be used at additional cost in computation.

Switch-level modeling techniques have also been used to analyze MOS circuits symbolically and to extract a functional description of the circuit [28,68,69,70]. However, when the size of a channel-connected subcircuit is relatively large, the symbolic representation could grow in size exponential in the number of variables; in this case the numeric method might be more suitable. In general, one may apply a combination of symbolic and numeric techniques in analyzing the circuit. Switch-level techniques have also been used to perform fault simulation [71-74] and automatic test generation [75,76]. They have also been extended to the analysis of digital bipolar circuits [77].

2.4 MIXED-MODE OR HYBRID SIMULATION

An ideal simulator for VLSI circuits would be one which has the speed and efficiency of digital or logic simulators while providing the accuracy and detail of an analog simulator. An attempt to achieve this is through mixed-mode or hybrid simulation. In many of the VLSI circuits the detail and accuracy provided by the analog simulators are not required for the entire circuit under investigation, but only for some *critical* areas of the circuit. This is particularly true of large digital circuits, where often a simple digital simulation (gate-level or switch-level) provides sufficient information about the performance of much of the circuit, while some parts, such as sense amplifiers in memory circuits or tightly coupled analog blocks, might require more detailed modeling and analysis.

By providing a range of models, from highly accurate and complex analog device models to much less accurate but greatly simplified gate-level or switch-level models, the circuit designer can reduce the simulation time significantly by choosing the computationally less expensive models whenever it is appropriate and possible. Another property of large circuits which may be exploited is their relative inactivity or latency. In a typical VLSI circuit, usually only less than 20% of the signals change values significantly at any one time-instant.

Hybrid analysis programs allow the designer to use a combination of analysis techniques and models, ranging from circuit and timing simulation to much cheaper digital simulation, in the same program. These simulators, such as SPLICE [13], DIANA [14], and SAMSON [15], have been observed to realize a one or two order of magnitude reduction in simulation time and substantially lower memory than standard circuit simulators, while still providing a detailed circuit-level analysis where necessary.

Mixed-mode or hybrid simulators, however, work well as long as only small, isolated sections of the circuit need to be simulated as analog circuits. Unfortunately, the partitioning of the circuit into sections which require analog simulation and those which do not is not fully automatic; some amount of human intervention is still required. Furthermore, trying to combine analog and digital models in a single program requires rather unsatisfactory approximations at the interfaces. For example, if the output of a section of logic gates is to be interfaced to an input of a section modeled as an analog circuit, a logic-to-voltage waveform conversion is required. This, of course, cannot be done with any accuracy, since much of the necessary information is lacking. The resultant outputs of the analog section must then be viewed somewhat skeptically. Similarly, certain states used in logic simulators, such as the unknown state X, or the high-impedance state **H**, do not represent a single voltage, and therefore, cannot be interfaced with an analog simulator. Therefore, unless great care is exercised, a hybrid simulator could end up providing the accuracy of a logic simulator at the speed of an analog simulator, rather than *vice versa*.

2.5 SWITCH-LEVEL TIMING SIMULATION

The problem of *switch-level timing simulation* of a digital circuit can be defined as follows :

Consider the analog waveform $V_n(t)$, $t \in [t_0, t_f]$ at a certain node n in a digital circuit and choose p—1 threshold values, ordered as $v_1 < v_2 < \cdots < v_{p-1}$. Define the *p-state digital equivalent* of

V_n to be

$$X_n(t) = x_i \text{ if } v_i < V_n(t) \leqslant v_{i+1} \qquad (2.1a)$$

where $x_0, x_1, \cdots, x_{p-1}$ are the p digital states and v_0 and v_p are the minimum and maximum values of the analog waveforms respectively. We also define

$$T_n = \{t_k : V_n(t_k) \in \{v_1, v_2, \cdots, v_{p-1}\}\}. \qquad (2.1b)$$

as the set of *threshold-crossing-times* of the analog waveform at node n in the circuit, or alternatively, the set of state transition times of its p-state digital equivalent. The aim of a switch-level timing simulator is to obtain the p-state digital equivalents X_n for each node n of interest to the user, with special emphasis on computing (or estimating) the threshold-crossing-times accurately. For brevity in notation, we shall use SLT to stand for *switch-level timing*, and so the threshold-crossing-times will be referred to as *SLT estimates*.

Since most VLSI circuits are primarily digital in nature, the circuit designer is very often satisfied in performing an SLT simulation in the design-verification process since this enables him to estimate the propagation-delays, speeds of computation, optimal clocking-rates, etc. The usefulness of an SLT simulator can be measured by considering two factors, namely, the *simulation cost*, which is primarily an increasing function of the CPU time and memory used, and secondly, the accuracy of the SLT estimates. There are two major approaches that could be used to perform an SLT simulation on a large digital circuit :

(1) Use an analog simulator and convert the resulting analog waveform into their p-state digital equivalents directly,

by choosing an appropriate set of p-1 threshold voltages.

(2) Use a digital simulator with delay estimation that computes the p-state digital waveform at each circuit node and generates the SLT estimates.

Since it is impossible to obtain the exact waveforms analytically, in a typical VLSI circuit, the SLT estimates produced by standard circuit simulators are considered accurate enough and are often taken as references to compare the accuracies of the SLT estimates produced with other simulators. Simulators using the first approach include the so-called *timing simulators* such as MOTIS [5,7], MOTIS-C [6], and PREMOS [8], which are analog simulators using relaxation techniques to speed up the simulation process.

In spite of the several attempts made to speed up the performance of standard circuit simulators, analog simulators are still very expensive to use to analyze circuits with more than 10k devices. Digital simulators, on the other hand, have a distinct advantage in speed over analog simulators. Several large circuits with over 100k transistors have been successfully handled by these simulators. However, they provide rather inaccurate SLT information due to the poor modeling of the dynamics of the circuits. Most digital simulators produce two-state digital waveforms, and account for the circuit dynamics by *delaying* the transition between states. In all cases the delays are taken to be *single-threshold* delays. Furthermore, these simulators do not take into account the dependence of propagation delays on circuit parameters, such as load capacitance, strengths of devices, input slew-rates, and other factors.

Based on the above facts, one can conclude that the circuit designer who wishes to use one of the existing analog or digital simulation tools to generate SLT estimates in VLSI circuits is placed in a difficult situation. Analog simulators provide fairly accurate SLT estimates at prohibitive simulation costs, while digital simulators can handle entire VLSI circuits but provide very poor SLT estimates, or sometimes, none at all.

It is therefore clearly necessary to provide the circuit designer with a simulation tool capable of providing accurate SLT estimates for VLSI circuits at reasonable simulation costs; thus, having the best features of both analog and digital simulators. To this end, one is more likely to succeed in trying to incorporate better timing models into digital simulators since efforts to speed up analog simulators seem to be approaching a limit which is far below the speeds of the digital ones. Restricting oneself to the MOS technology seems to make the problem a little easier. An attempt has been made to model the MOS transistor as a linear resistor resulting in an RC-delay model for the circuit dynamics which is used in RSIM [29]. This is a logic-level timing simulator which predicts the logic state of a node and uses an RC time-constant to estimate the transition times if the node changes state. The transistor model in RSIM is a gate-voltage dependent resistance R_{ds} between drain and source terminals. When the switch is *closed*, we have $R_{ds} = R_{eff}$, when *open* $R_{ds} = \infty$, and when in the *unknown* state (which means $v_{gate} = X$) the drain-source connection is described by a *resistance interval*, i.e., $R_{ds} = [R_{eff}, \infty]$. The effective resistance R_{eff} is determined separately for each transistor as a function of the device width and length, the transistor type, and other device parameters. The determination of the effective resistance is made once for each transistor and is about the only device information used by RSIM. Voltages in the RSIM model are quantized into one of three values, 0,1,X, and decided by choosing two threshold voltages, v_{low} and v_{high}.

The effect of the resistive network on a particular node is modeled by a Thevenin equivalent circuit. The values of V_{thev} and R_{thev} are computed, in some cases approximately, based on a series-parallel-type approach which is illustrated in [30]. The value of V_{thev} (which may be a voltage interval in some cases) decides the new state of the node. If the new value at a node is different from the previous one, then a transition is scheduled $R_{thev}C_{load}$ time-units later, where C_{load} is the net capacitance at the node. Actually, RSIM uses three values of the effective resistance for a transistor, namely, a static value used to determine V_{thev}, and two others to be used in determining rise and fall delays. All these values are determined in a presimulation phase using an accurate circuit simulator such as SPICE2 [1]. Charge sharing effects are also taken into account. A nice feature of this type of simulation is that the X state does not impose any particular difficulty as far as the simulation is concerned. The simulator is event-driven and is fast enough to simulate circuits of up to 50k transistors. The SLT estimates are, however, computed only by single threshold RC delays and are sometimes found to be even more than 30% off when compared with those of SPICE2, especially in the case of MOS circuits with large pass-transistor chains.

CHAPTER 3

MOS NETWORK PARTITIONING AND ORDERING

In this chapter an MOS network model that can be used to perform switch-level simulation is presented. The entire MOS network is first partitioned into several subnetworks, or blocks. The transistors within a block may be further partitioned into driver and pass transistors if so desired. The set of blocks is then partitioned into its strongly connected components (SCC). The SCCs in the network are then ordered for simulation.

3.1 MOS NETWORK COMPONENTS AND MODELS

An MOS digital network Ω consists of a set of *nodes* N interconnected by a set of MOS transistors M. The network description can be extracted directly from the layout using computer programs known as circuit extractors [61,62], or has to be given by the user. In any case, the network description is assumed to contain a netlist of all the MOS transistors along with several geometrical and process parameters, such as length (L) and width (W) of each device, zero-bias device threshold voltage (V_t), transconductance parameter (K_p), the type of device (i.e., n-channel or p-channel), the analog waveform at the input sources, and a lumped capacitance **CAP(n)** from each node $n \in N$ to ground. Specifying a grounded capacitance from each node might seem to be a restriction, but most circuit extractors could be asked to compute equivalent device capacitances along with the capacitance due to the interconnect regions.

In switch-level modeling, there are two types of nodes, namely,

(1) **Input nodes**, which are modeled as voltage sources and provide the strongest signals to the network from the outside. Examples of input nodes include the power supply (V_{DD}), the ground node (GND), as well as all the input clock signals.

(2) **Normal nodes** or **Storage nodes**, which are the remaining nodes in the circuit. These are the weakest nodes as they cannot force their signals on a stronger node but are capable of storing a signal dynamically. These nodes, in turn, can be further subdivided into several strength classes depending upon their relative capacitance values.

In some simulators for circuits using n-channel MOS (NMOS) technology a third type of node, namely, a **pullup node**, is defined as a node connected to the power supply V_{DD} via a pullup resistor. A pullup node retains the value of the supply unless forced to ground through a path of conducting devices. The output node of a Boolean logic gate realized using NMOS technology with depletion loads is an example of a pullup node. A pullup node is *weaker* than an input node but is *stronger* than a storage node.

An MOS transistor is modeled as a three-terminal device with a switch between the drain and source terminals and the signal at the gate controlling the status of the switch. We will only consider transistors whose drain and source nodes are different. In some technologies the drain and source regions of a transistor might correspond to the same net in the layout as a means of implementing a capacitance. We shall, however, assume that the circuit extraction program used identifies this situation correctly as a capacitor rather than a transistor. Associated with each device $\mathbf{m} \in \mathbf{M}$ is a resistance $\mathbf{RES(m)} = \frac{2L}{K_p W}$.

There are three types of MOS transistors, namely,

(1) n-channel, enhancement type (**n—enh**),

(2) p-channel, enhancement type (**p—enh**), and

(3) n-channel, depletion type (**n—dep**).

The **n—enh** devices are characterized by positive device threshold voltages (i.e., $V_t > 0$) and behave as voltage controlled switches. Similarly, **p—enh** devices have a negative V_t and also behave as voltage controlled switches. The **n—dep** devices, on the other hand, are n-channel transistors with a negative V_t and are mainly used to implement pullup resistors. The gate and source nodes of an **n—dep** device are usually shorted resulting in a two-terminal resistor.

Mathematically, the MOS network $\Omega(N,M)$ can be specified by giving a listing of nodes in N and transistors in M and the functions specified in Table 3.1.

Table 3.1: A list of functions used to describe an MOS network

NODTYP :	N→{ input, pullup, normal }	node type
TRNTYP :	M→{ n–enh, p–enh, n–dep }	transistor type
GATE :	M→N	gate node
SOURCE :	M→N	source node
DRAIN :	M→N	drain node
CAP :	N→[C_{min} , C_{max}]	node capacitance
RES :	M→[R_{min} , R_{max}]	transistor resistance

3.2 PARTITIONING THE MOS NETWORK INTO BLOCKS

In this section we describe the strategy and algorithms to partition the MOS network $\Omega(N,M)$ into several transistor-disjoint subnetworks $\Omega_1, \Omega_2, \cdots, \Omega_s$, where each subnetwork or block Ω_i has a certain special configuration that would aid the simulation process. The partitioning strategy is basically to group together a set of transistors to constitute a subnetwork or a block if they have a common-channel-path between their source and drain nodes (a notion that will be made precise, later in Section 3.2.2). The key notion used in the partitioning algorithm is that of *input-node-splitting* which will also be defined in Section 3.2.2. It is much easier to formally present our ideas and concepts if the network is viewed as an undirected graph; therefore, we begin by reviewing some basic fundamentals from graph theory for the sake of completeness and also for the benefit of readers who are not familiar with the subject. An excellent reference on the fundamentals of graph theory is a book by Bondy and Murty [55]. The readers already familiar with the basic concepts may skip to Section 3.2.2 without loss of continuity.

3.2.1 Review of Graph theory

An *undirected graph* H is an ordered triple $(V(H), E(H), \psi_H)$ consisting of a non-empty set $V(H)$ of *vertices*, a set $E(H)$ of *edges*, disjoint from $V(H)$, and an *incidence function* ψ_H which associates with each edge of H an unordered pair of (not necessarily distinct) vertices in H. If e is an edge and v and w are vertices such that $\psi_H(e) = \langle v, w \rangle$, then e is said to *join* v and w, the vertices v and w are called the *ends* of e and further, v and w are said to be *adjacent* in H. In this case we will usually refer to the edge e as simply $\langle v, w \rangle$. The set of all vertices in H that are adjacent to the vertex v is denoted by $Adj_H(v)$. The two ends of an edge are incident with the edge and *vice versa*. If the two ends of an edge are the same then the edge is called a *loop*; otherwise, it is a *link*. The symbols $\nu(H)$ and $\epsilon(H)$ are used to denote the number of vertices and edges in graph H respectively, i.e., $\nu(H) = |V(H)|$ and $\epsilon(H) = |E(H)|$. When only one graph is under discussion it will be denoted by H, and we will use V, E, ν and ϵ instead of $V(H)$, $E(H)$, $\nu(H)$ and $\epsilon(H)$. An undirected graph is usually represented pictorially on a plane by associating one point (or a dot) for each vertex and joining two points by a line (not necessarily straight) if the corresponding vertices are joined by an edge. As an example, consider a graph H with

$$V(H) = \{v_1, v_2, v_3, v_4, v_5\}$$

$$E(H) = \{e_1, e_2, e_3, e_4, e_5, e_6\}$$

and the incidence function defined by

$$\psi_H(e_1) = \langle v_1, v_2 \rangle,\ \psi_H(e_2) = \langle v_2, v_4 \rangle,\ \psi_H(e_2) = \langle v_4, v_3 \rangle$$

$$\psi_H(e_4) = \langle v_3, v_1 \rangle,\ \psi_H(e_5) = \langle v_2, v_2 \rangle,\ \psi_H(e_6) = \langle v_1, v_4 \rangle.$$

The pictorial representation of this graph is shown in Figure 3.1. The point representing the vertex v_5 is isolated in the picture since there are no edges incident on this vertex in this case. Hence, vertices with no edges incident on them are called *isolated* vertices. Henceforth, we shall refer to a graph by its pictorial representation.

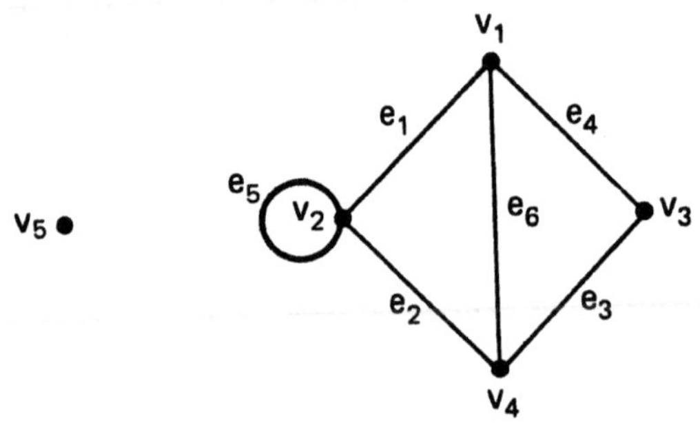

Figure 3.1 : An undirected graph H

A graph F is a *subgraph* of H, written as $F \subseteq H$, if $V(F) \subseteq V(H)$, $E(F) \subseteq E(H)$, and ψ_F is a *restriction* of ψ_H to $E(F)$. If V' is a subset of V, then the subgraph of H whose vertex set is V' and whose edge set is the set of *all edges* of H that have both ends in V' is called the *induced subgraph* of H by V' and is denoted by $H[V']$. The induced subgraph $H[V \backslash V']$, denoted by $H-V'$, is the subgraph obtained from H by deleting the vertices from V' along with all their incident edges. If E' is a nonempty subset of E, then the subgraph of H *induced* by E' is the one with vertex set as the set of the ends of edges in E' and edge set E' and is denoted by $H[E']$. The subgraph obtained from H by deleting the edges in E' is denoted as $H-E'$. It must be pointed out that deleting vertices from a graph also involves deleting incident edges; however, deleting edges involves only the removal of edges while leaving the set of vertices intact, i.e., $V(H-E') = V(H)$. Similarly, $H+E'$ is a graph obtained from H by inserting a new set of edges E' which are disjoint from the old set of edges $E(H)$. Again, in this case, the ends of the edges in E' must necessarily be in V(H) since no new vertices are added. If F and H are two undirected graphs then their *union* is a graph, denoted by $F \cup H$, whose vertex set is $V(F) \cup V(H)$ and whose edge set is $E(F) \cup E(H)$. If F and H are *disjoint* graphs, then their union is usually denoted by $F+H$. The *degree* $d_H(v)$ of a vertex v in H is the number of edges incident on v, with each loop counting as two edges.

A *walk* in an undirected graph H is a finite, nonempty sequence $W = v_0 e_1 v_1 e_2 v_2 \cdots e_k v_k$ whose terms are alternately vertices and edges in H such that for each $1 \leqslant i \leqslant k$, the ends of e_i are v_{i-1} and v_i. In this case W is said to be a walk from v_0 to v_k, or a (v_0,v_k)-walk in H, and the integer k is called the *length* of the walk. The vertices v_0 and v_k are called the *origin* and *terminus* of the walk respectively, while the vertices $v_1,v_2,\ldots,v_{k-1}$ are its *internal* vertices. If all the vertices in a walk are distinct then it is said to be a *path*. Usually, the subgraph of H whose vertices and edges are terms of a path is also referred to as a path. A walk is closed if it has positive length (i.e., $k>0$) and its origin and terminus are the same. A closed walk whose origin and internal vertices are distinct is a *cycle*. Just as with paths we sometimes use the term "cycle" to denote the graph corresponding to the cycle. Two vertices v and w of H are said to be *connected* if there exists a (v,w)-path in H. A subgraph F is a *component* of H if it is a maximal induced subgraph such that any two of its vertices are connected. If H has only one component then H is *connected;* otherwise, it is *disconnected*. The number of components of H is denoted by $\omega(H)$.

3.2.2 Blocks of an MOS Network

We are now ready to represent an MOS network by an undirected graph. An undirected graph H represents a network Ω if there is a vertex in H corresponding to each node in the network and an edge between two vertices if the corresponding nodes are the source and drain nodes of some MOS transistor. More formally,

Definition 3.1 : An undirected graph $H(V,E,\psi_H)$ is said to *represent* an MOS network $\Omega(N,M)$ if there exist bijections

$\theta : V \rightarrow \mathbf{N}$ and $\phi : E \rightarrow \mathbf{M}$ such that $\psi_H(e) = <v, w>$ if and only if $\{\theta(v), \theta(w)\} = \{\mathbf{DRAIN}(\phi(e)), \mathbf{SOURCE}(\phi(e))\}$.

A vertex v in a graph is said to be *loop-free* if there are no loops incident on v. The entire graph is loop-free if all its vertices are loop-free, i.e., it has no loops as edges. A graph is said to be *isolated* if all its vertices are isolated, i.e., it has no edges.

Theorem 3.1 : If H represents an MOS network Ω, then H is a loop-free graph.
Proof : If e is a loop in H then it follows from the above definition that $\theta(v) = \mathbf{DRAIN}(\phi(e)) = \mathbf{SOURCE}(\phi(e))$ where v is the vertex incident with the loop. But this is impossible since this means that there is some transistor in the network with source and drain nodes tied together and we do not consider such networks in our model as explained in Section 3.1. Hence, H has no loops. □

Let $\mathbf{N_I}$ and $\mathbf{N_N}$ denote the sets of input and normal nodes in the network, respectively. It must be noted that, by definition, the ground node (**GND**) is treated as an input node. Let

$$\mathbf{M_{enh}} = \{\mathbf{m} \in \mathbf{M} : \mathbf{TRNTYP(m)} \in \{\mathbf{n-enh}, \mathbf{p-enh}\}\} \quad (3.1)$$

denote the set of enhancement transistors in the network. Let $\mathbf{N_O} \subseteq \mathbf{N_N}$ be the subset of normal or storage nodes at which the user wishes to observe the output waveforms. Also, let

$$\mathbf{N_G} = \{\mathbf{n} \in \mathbf{N_N} : \mathbf{n{=}GATE(m)} \text{ for some } \mathbf{m} \in \mathbf{M_{enh}}\} \quad (3.2)$$

denote the set of normal nodes that are gate nodes of the enhancement transistors in the network.

We now introduce the notion of splitting a vertex in a graph. Consider an undirected graph $H(V, E, \psi_H)$ and vertex v in the graph of degree $k \geqslant 1$, i.e. $d_H(v) = k \geqslant 1$.

Definition 3.2 : Let v be a loop-free vertex of degree $k \geqslant 1$ in an undirected graph H. The *v-split graph* or the graph obtained on splitting v in H, is a graph obtained by splitting the vertex v into k new vertices $y_1, y_2, \cdots, y_k$, with each edge formerly joining the vertex v to w_i now joining y_i to w_i. We denote the v-split graph as $H \bullet v$. Thus, *splitting* a vertex creates a new graph with $k-1$ more vertices but with the same set of edges. This is in contrast to the notion of *adding* new edges to a graph in which case the vertex set is unaltered while new edges are added to the graph. It can easily be seen that if $k=1$, then splitting the vertex v does not alter the graph, i.e., $H \bullet v = H$ if $d_H(v)=1$. Similarly, the notion of vertex splitting can be extended to include the case $k=0$ by defining $H \bullet v = H$ if $d_H(v)=0$. If $V' = \{v_1, v_2, \cdots, v_q\}$ is a subset of loop-free vertices in H then the V'-split graph of H can be defined as follows :

$$H \bullet V' = (\cdots ((H \bullet v_1) \bullet v_2) \cdots) \bullet v_q \tag{3.3}$$

The graph $H \bullet V'$ is well-defined since the order in which the vertices of V' are split does not matter. As an example consider the graph shown in Figure 3.2(a). The graph obtained by splitting the vertices v_1 and v_2 is shown in Figure 3.2(b).

We will now partition the MOS network into blocks.

a)

b)

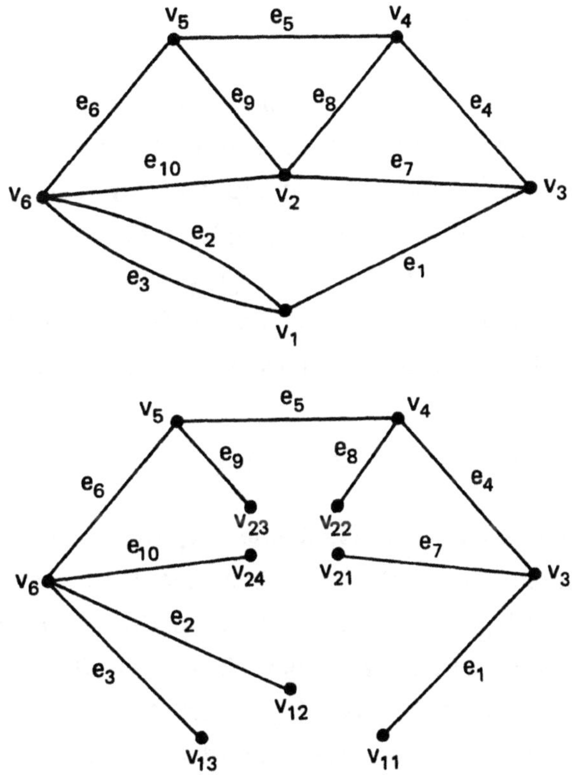

Figure 3.2(a) : A loop-free graph H
(b) : The graph obtained by splitting v_1 and v_2 in H

Instead of dealing with the network $\Omega(N,M)$ we will be concerned with the graph $H(V,E)$ that represents the network. Obtaining the graph that represents the network merely involves altering the data structure that represents the network to the one that represents a graph.

Definition 3.3 : Let the graph H represent the network Ω and let V_I denote the set of input-vertices in the graph that correspond to the nodes of input strength N_I in the network. Let $H_I = H \bullet V_I$ denote the graph obtained from H by splitting all the vertices in V_I. Let $C_1, C_2, \cdots, C_s$ be the various connected components of H_I. A *block* Ω_i is defined to be the subnetwork of transistors corresponding to the edges of component C_i, i.e.,

$$\Omega_i = \{m \in M : \phi^{-1}(m) \in C_i\}. \tag{3.4}$$

For each block Ω_i define N_{DS}^i to be the set of drain and source nodes of all the enhancement transistors within Ω_i. Also, let N_G^i denote the set of gate nodes of all the enhancement transistors within Ω_i. Then each block Ω_i has a set of *input nodes*, denoted by $INP(\Omega_i)$, defined to be the set of circuit nodes which are the gate nodes of some enhancement transistor within Ω_i together with the drain and source nodes of enhancement transistors in Ω_i of input strength excluding V_{DD} and GND, i.e.,

$$INP(\Omega_i) = N_G^i \cup [N_{DS}^i \cap (N_I - \{V_{DD}, GND\})] \tag{3.5}$$

Similarly, each block has a set of *output nodes*, denoted by $OUT(\Omega_i)$, defined to be the set of circuit nodes of normal strength that are drain or source nodes of some transistor within Ω_i and gate nodes of some transistor outside Ω_i or a user-defined output node, i.e.,

$$\mathrm{OUT}(\Omega_i) = N_N \cap [N_{DS}^i \cap (N_G \cup N_O)]. \quad (3.6)$$

3.2.3 Partitioning Algorithm and Its Complexity

We now outline the algorithm that partitions an MOS network into blocks. The algorithm begins by first splitting all the input vertices V_I in the graph representing the MOS network. It then identifies the various connected components of this new graph. The transistors corresponding to a connected component constitute a *block* in the original network Ω.

Algorithm 3.1 : Partitioning into Blocks

```
procedure PARTITION (H)
begin
        H_I ← H;
        for each v ∈ V_I do
                H_I ← SPLIT(H_I , v);
        end for
        Φ ← COMPONENT(H_I );
        return (Φ)
end
```

In the above algorithm we must ensure that any vertex that is split in a graph is, in fact, loop-free. This is, indeed, the case since from Theorem 3.1 we have that the entire graph H is loop-free. In order to discuss the details of procedures SPLIT and COMPONENT, used in the above algorithm, we will assume that the reader is familiar with the *adjacency list* [56] representation for graphs.

```
procedure SPLIT (F , v )
begin
        k ←d_F (v );
        if (k = 1) return (F )
        LOC ←FIRST_EDGE(v );
        FIRST_EDGE (v )←0;
        REMOVE (v );
        for each i = 1 to k do
                ADD_VERTEX(v_i );
                FIRST_EDGE(v_i )←LOC ;
                LOC 1←NEXT_EDGE(LOC );
                ADJACENT(LINK(LOC ))←v_i ;
                NEXT_EDGE(LOC )←0;
                LOC ←LOC 1;
        end for
        return (F )
end
```

The procedure COMPONENT (F), described below, returns the various components in the graph F. A Boolean array of the vertices is maintained to mark a vertex as **new** or **old**, such that every time this array is altered, a pointer, LFN, exists to indicate the location of the first vertex marked **new**. Initially all vertices of F are marked **new**. The procedure begins by starting from the first vertex marked **new** and using a breadth-first search algorithm [56] to determine all the vertices connected to the starting vertex *via* a path in F. These vertices induce a component and are all marked **old**. The whole process is repeated by starting from the first vertex that is now still marked **new** until all vertices are marked **old**, indicated by $LFN=0$.

```
procedure COMPONENT (F)
begin
        i ←0;
        mark all vertices as new;
        initialize STACK to empty;
        repeat
                i ←i +1;
                C_i ←Ø;
                v ←VERTEX(LFN );
                MARK[v ]← old;
                push v into STACK;
                while STACK is not empty do
                        pop w from STACK;
                        LOC ←FIRST_EDGE(w );
                        while (LOC ≠ 0) do
                                e ←LABEL(LOC );
                                C_i ←C_i ∪{e };
                                x ←ADJACENT(LOC );
                                if MARK [x ] = new then
                                        push x into STACK;
                                        MARK [x ]← old;
                                end if
                                LOC ←NEXT_EDGE(LOC );
                        end while
                end while
        until LFN = 0;
        return {C_1,C_2, · · · ,C_i }
end
```

The *time-complexity* of an algorithm to solve a problem is said to be $O(f(n))$ if the maximum amount of computation time (or number of computation steps) taken by the algorithm is at most $cf(n)$ over all inputs of size n, where c is some constant. The space-complexity is similarly defined as an upper bound on the amount of space required by an algorithm to solve a problem. Two excellent references on the subject of time- and space-complexity of algorithms are Aho, Hopcroft, and Ullman [56] and Garey and Johnson [57]. In most graph algorithms the input size n is taken to be $|V|+|E|$, where $|V|$ and $|E|$ are the number of vertices and edges in the graph respectively. The time- (or space-) complexity is said to be *linear* if $f(n)=n$. The following theorem demonstrates that Algorithm 3.1, described above, is of linear time-complexity.

Theorem 3.2 : The Algorithm 3.1, described above, partitions the edges of H and its time-complexity is $O(|V|+|E|)$ where V is the set of vertices and E is the set of edges in graph H.

Proof : Consider the operation of splitting a vertex v of degree k from a graph F. This merely involves altering the data structure to represent the new graph and can be easily shown to have a time-complexity of $O(k)$. This procedure is repeated $|V_I|$ times in Algorithm 3.1; hence, the time-complexity of splitting all vertices of input strength from H is $O(q)$ where $q = \sum_{v \in V_I} d_H(v)$. However, $q \leq 2|E|$, since splitting vertices from a graph does not alter the edge-set of the graph. Therefore, the required splitting can be performed with a time-complexity of $O(|E|)$.

Now consider the outermost **repeat**-loop in procedure COMPONENT (F). Each time the algorithm enters this loop it produces a list of edges in a component of the graph F in computation time linearly proportional to the number of edges in that component [56]. Thus, if one does not have to scan the array to look for a starting vertex marked **new**, which is possible by maintaining the required pointer, LFN, the time-complexity of the entire procedure COMPONENT (F) without accounting for the time required to update the pointer LFN is $O(|E(F)|)$. Clearly, the time required to initially mark all vertices as **new** and subsequently update LFN is $O(|V(F)|)$. Therefore, the total time-complexity of COMPONENT (F) is $O(|V(F)|+|E(F)|)$. But $|V(H_I)| \leqslant |V(H)|+2|E(H)|$ and $|E(H_I)|=|E(H)|$; hence, COMPONENT (H_I) will take $O(|V(H)|+|E(H)|)$ computation time. This concludes the proof. □

To model the voltage-source elements connected to the input nodes of the network we introduce a new type of block called input sources (SRC) consisting of only a node of input strength (and no transistors). This node is said to be the *output node* of the SRC.

In this section, thus far, we have shown how we partition an MOS network $\Omega(N,M)$ into several transistor-disjoint subnetworks called blocks. We have also demonstrated an algorithm by which this partitioning can be achieved in computation time that is at most linearly proportional to the number of nodes and transistors in the network. We will use the symbol Σ to denote the set of partitioned-blocks in the network, and henceforth, we shall refer to the partitioned MOS network as $\Omega(N,M,\Sigma)$. Furthermore, INP(Ω_i) and OUT(Ω_i) will be used to denote the sets of input and output nodes of subnetwork $\Omega_i \in \Sigma$.

3.2.4 A CMOS Example

In this section we will consider a complementary MOS (CMOS) network shown in Figure 3.3 as an example to illustrate the partitioning algorithms described earlier. This network consists of 12 nodes and 14 transistors. Node n_0 is GND while node n_1 is V_{DD}. The nodes

$$N_I=\{n_0,n_1,n_2,n_3,n_4,n_5\},$$

are the nodes of input strength, while the nodes

$$N_N=\{n_6,n_7,n_8,n_9,n_{10},n_{11}\},$$

are of normal strength. The transistors

$$M_n=\{m_{11},m_{12},\ldots,m_{17}\}$$

are the n-type enhancement devices while

$$M_p=\{m_{21},m_{22},\ldots,m_{27}\}$$

are the p-type enhancement devices. There are no pullup nodes nor depletion devices in this circuit.

The graph H representing this network is shown in Figure 3.4. We only show the non-isolated vertices in the graph. Also, we refer to the vertices and edges of the graph as nodes and transistors in the network, respectively, for the sake of convenience, i.e., in this case the bijections θ and ϕ used in Definition 3.1 are both identity mappings. The graph H_I obtained by splitting the nodes of input strength from H is shown in Figure 3.5.

The graph H_I has 3 components. Hence, the network has three blocks, namely,

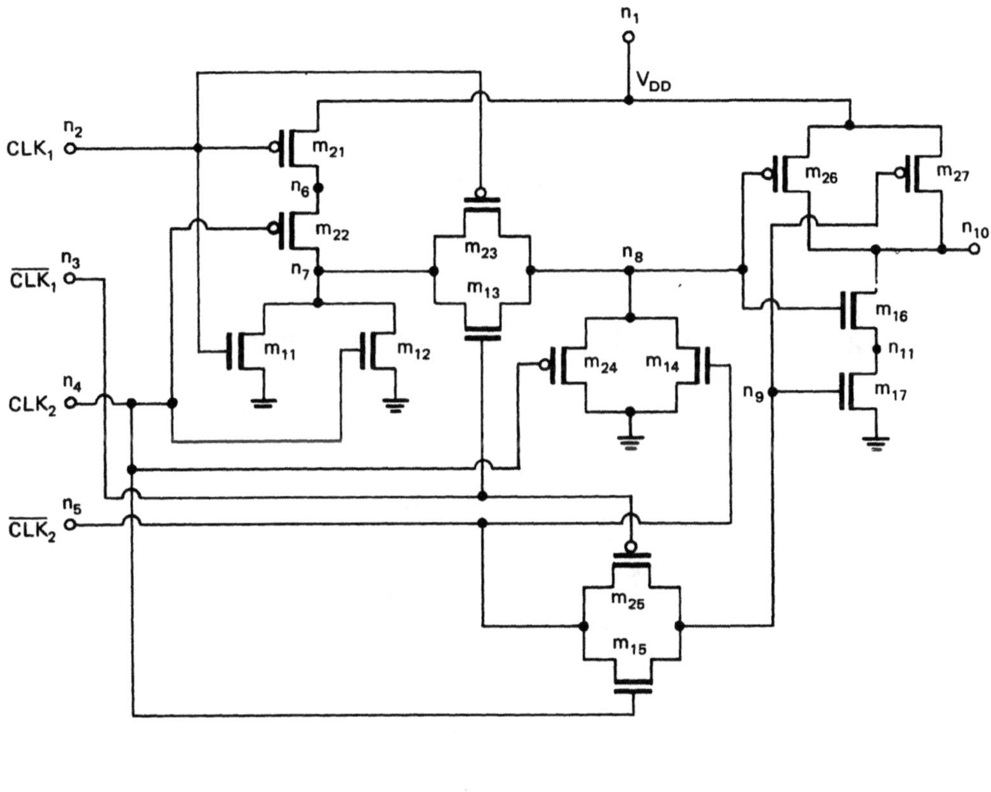

Figure 3.3 : An example of a CMOS network

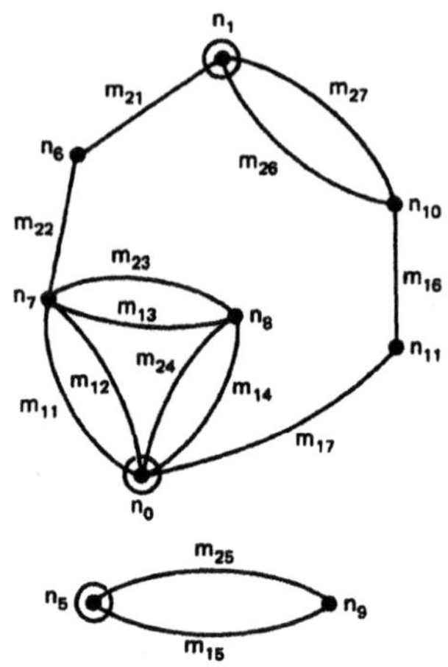

Figure 3.4 : The graph H representing the network in Figure 3.3

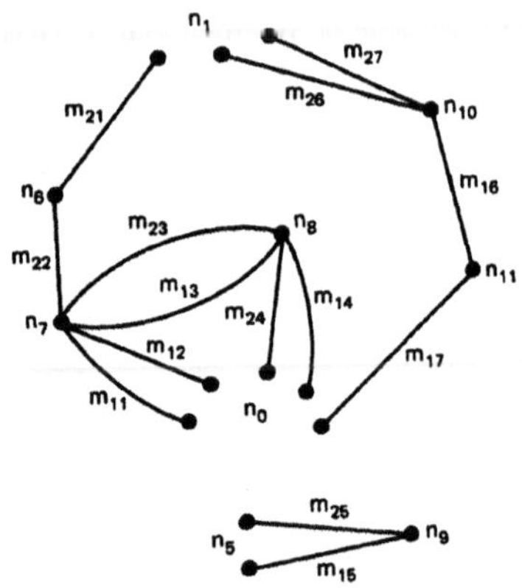

Figure 3.5 : The graph H_I for H in Figure 3.4

$$\Omega_1=\{m_{11},m_{12},m_{13},m_{14},m_{21},m_{22},m_{23},m_{24}\},$$

$$\Omega_2=\{m_{15},m_{25}\},\ \text{and}\ \Omega_3=\{m_{16},m_{17},m_{26},m_{27}\}.$$

Table 3.2 contains a listing of the set of its input and output nodes in each block. In addition, the network also has 5 SRCs which are listed in Table 3.3 along with the node of input strength in each of them.

Table 3.2: A list of blocks of the network in Figure 3.3

Block	Input Nodes	Output Nodes
Ω_1	n_2,n_3,n_4,n_5	n_8
Ω_2	n_3,n_4,n_5	n_9
Ω_3	n_8,n_9	n_{10}

Table 3.3: A list of SRCs of the network in Figure 3.3

Block	Output Node
SRC_1	n_1
SRC_2	n_2
SRC_3	n_3
SRC_4	n_4
SRC_5	n_5

3.3* PARTITIONING INTO DRIVER AND PASS TRANSISTORS

In the previous section we described a linear time algorithm to partition the transistors of an MOS network into several channel-connected components, called blocks. These transistor-blocks are also referred to as transistor-groups in a switch-level simulator such as MOSSIM [23]. Such a partitioning is adequate for switch-level logic simulation and is used in a variety of switch-level simulators [23-28, 64-70]. In a switch-level timing simulator such as MOSTIM [32], however, the set of MOS transistors within a block or group have to be further partitioned into *driver* transistors and *pass* transistors due to the following reasons.

(1) The channel-connected components consisting of only driver transistors, which will be referred to as *multi-functional blocks* (MFB), behave as unidirectional multi-input single output functional units similar to Boolean logic gates. The signals at the nodes internal to an MFB need not be evaluated during simulation since they do not interact with the rest of the network and are of no interest as such to the user.

(2) A pass transistor, however, is a bidirectional device and essentially causes problems in simulating *pass transistor blocks* (PTB) with the Boolean gate model. The full power of conventional switch-level simulation [24] is actually required to correctly verify the logic in a PTB.

* The reader may wish to skip this section initially and move on to Section 3.4 without loss of continuity.

(3) The delay characteristics of MFBs are very different from those of PTBs. As far as I/O-propagational delay is concerned, an MFB can be treated as an equivalent simple inverter using simple impedance scaling techniques. The knowledge of the delay characteristics of a simple inverter is, therefore, sufficient to make a fairly accurate prediction of the delay through any general MFB.

A collection of driver transistors in an MFB constitutes a two-terminal switching network between the output node and the ground node. The signals at the internal nodes of such sub-networks do not explicitly affect the rest of the network and need not be computed as such. Furthermore, the capacitances at these internal nodes tend to be much smaller than gate capacitances and can either be ignored or lumped at the output node of the MFB without any severe loss of accuracy in delay computation. An MFB can, therefore, be simulated using relatively simpler techniques for both switch-level simulation as well as delay computation.

In case of a collection of pass transistors, however, the situation is entirely different. First of all, the pass transistors in a PTB constitute a multi-port, multi-terminal switching network. Hence, the full power of conventional switch-level simulation is required to even generate zero-delay waveforms for a PTB. Secondly, a typical PTB consists of several nodes with a wide range of node capacitances. This requires more complex methods involving computation of Elmore time-constants [32,45,46] to provide accurate timing. Finally, pass transistors are often used to realize dynamic memories, transferring information to and from busses etc.; hence, situations such as charge sharing and conflict between two or more strong signals often arise during simulation of PTBs. A special technique has to be used to compute the delays in these cases.

It is, therefore, clear that in order to obtain fast and accurate timing information through switch-level simulation of a very large MOS integrated circuit, it is imperative to separate driver transistors from pass transistors. Identification of pass transistors has also been considered in PRIMO [60], which is a circuit partitioner for SAMSON [15]. The approach used in PRIMO is based on a recursive depth first traversal technique which might also require back-tracking. Furthermore, the emphasis is on the detection of so-called *bidirectional* pass transistors which are considered different from the unidirectional transistors. For example, consider a network consisting of an NMOS inverter driving a pass transistor as shown in Figure 3.6. Suppose the output of the pass transistor is connected to the gate terminal of some other transistor. It is clear that while the output of the inverter is a Boolean function of its input the output of the pass transistor is in fact not so. However, the pass transistor, if turned ON, merely transmits information from the inverter's output to its own output and is, thus, unidirectional in this sense. Hence, PRIMO treats both the inverter and the pass transistor as a single logic block. The outputs of these logic blocks, however, may not always be Boolean functions of their inputs.

In a circuit partitioner for a switch-level simulator such as MOSTIM, the emphasis is on the detection of all pass transistors so that on their removal the network consists of simply

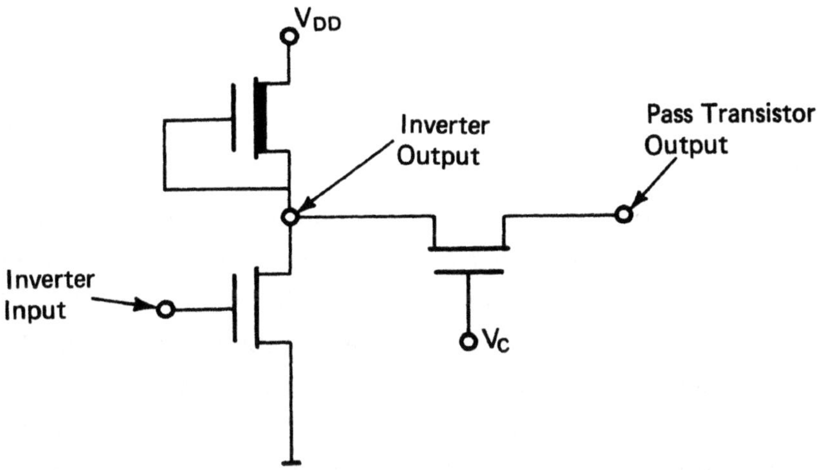

Figure 3.6 : An NMOS inverter driving a pass transistor

single-output combinatorial Boolean logic gates. The objective of the partitioning algorithm presented in this section is, therefore, quite different for that used in PRIMO. Furthermore, our approach is more formal, precise and general, and is not recursive; thus, resulting in a linear time implementation. Recognizing pass transistors is also useful in some of the new approaches to circuit/timing simulation [7,15].

In this section the algorithms for separating the driver transistors from pass transistors will be outlined and discussed for NMOS circuits with depletion loads only. Several extensions to handle circuits with other technologies, such as CMOS, will, however, be mentioned briefly at the end of this section.

The reader may skip the remainder of this section, and move on to Section 3.4, without loss of continuity, if distinction between driver and pass transistors is not a critical issue.

3.3.1 Motivation

Before going into the formal definitions, we would like to provide the reader with some intuition on deciding between driver and pass transistors in a network. We define *external* nodes to be the set of nodes of input strength apart from the ground node together with those nodes of normal strength that are either gate nodes of enhancement transistors or are user-requested output nodes. Now consider a graph on the nodes of an NMOS network with an edge between the drain and source nodes of each enhancement transistor. Let us focus our attention on a pullup node, say $\mathbf{n_p}$ in the graph. For each such pullup node we consider the subnetwork composed of the depletion device connected to the pullup node and the transistors corresponding to all the paths between $\mathbf{n_p}$ and the ground node. If all the nodes corresponding to the internal vertices in each of these paths are of normal strength and if none of these nodes is an external node, we can then define the above subnetwork to be a multi-functional block (MFB) and all the enhancement transistors in it as driver transistors. Furthermore, each MFB must contain a unique pullup node.

Consider an example of an NMOS network shown in Figure 3.7(a) and the corresponding graph in Figure 3.7(b). From the above definition, clearly $\mathbf{m_3}$ is a driver transistor. The

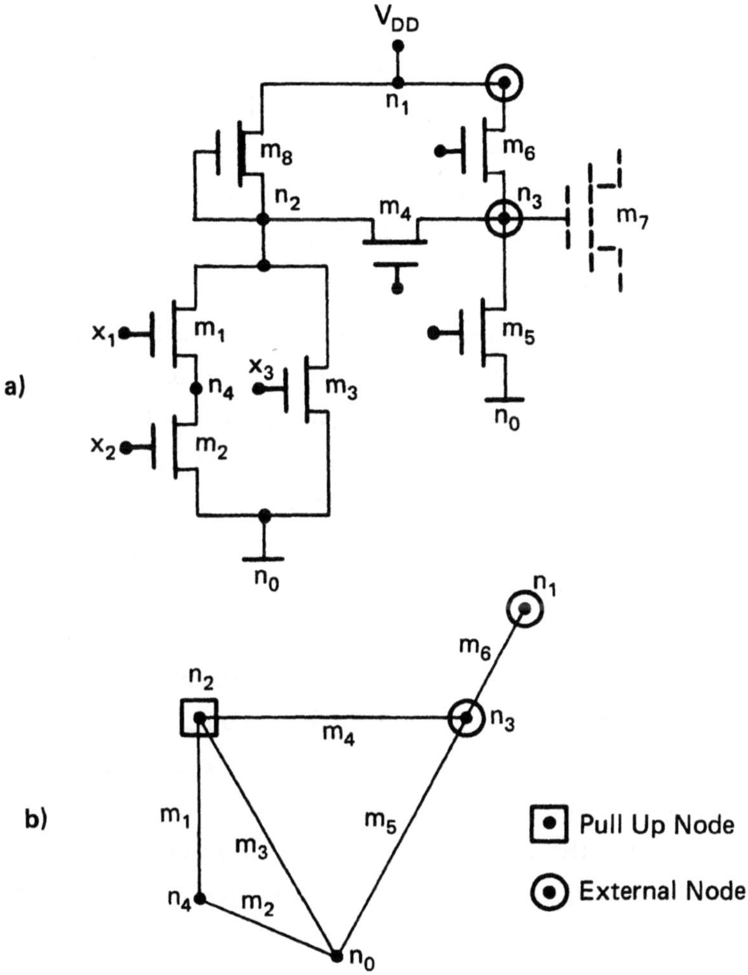

Figure 3.7(a) : An NMOS circuit with external nodes
(b) : The graph representing the circuit in part (a)

transistors $\mathbf{m}_1$ and $\mathbf{m}_2$ are also drivers since the internal node, $\mathbf{n}_4$, is of normal strength and is not an external node. The node $\mathbf{n}_3$ is an external node by definition; hence, $\mathbf{m}_4$ and $\mathbf{m}_5$ are not driver transistors. In fact $\mathbf{m}_4$, $\mathbf{m}_5$, and $\mathbf{m}_6$ are pass transistors. The MFB corresponding to the pullup node, $\mathbf{n}_2$, in this example, is the subnetwork consisting of the depletion transistor $\mathbf{m}_8$ along with the driver transistors $\mathbf{m}_1$, $\mathbf{m}_2$, and $\mathbf{m}_3$. The subnetwork composed of the pass transistors $\mathbf{m}_4$, $\mathbf{m}_5$, and $\mathbf{m}_6$ is called a pass transistor block (PTB).

The above definition for a driver transistor is, in fact, only a sufficient condition satisfied by driver transistors as the following example demonstrates. Consider the NMOS network shown in Figure 3.8(a), and the corresponding graph in Figure 3.8(b). In this example $\mathbf{n}_4$ is an external node by definition. Let us suppose $\mathbf{n}_3$ is simply a node of normal strength and is not an external node. In this case the path consisting of $\mathbf{m}_4$ and $\mathbf{m}_5$ would satisfy the above definition of driver transistors; hence, these transistors would be included in the MFB with pullup node $\mathbf{n}_2$. However, one needs to compute the signal at $\mathbf{n}_4$ since this determines the switching state of transistor $\mathbf{m}_7$, and in order to do this, we need to compute the signal at node $\mathbf{n}_3$ which, by the above definition, is an internal node of an MFB. We, therefore, have to modify our definition of a driver transistor. To this end, we introduce the concept of a pseudo-external node.

A node of normal strength is said to be a *pseudo-external* node if it can be connected to an external node by a path that does not contain a pullup node or the ground node. Clearly, the

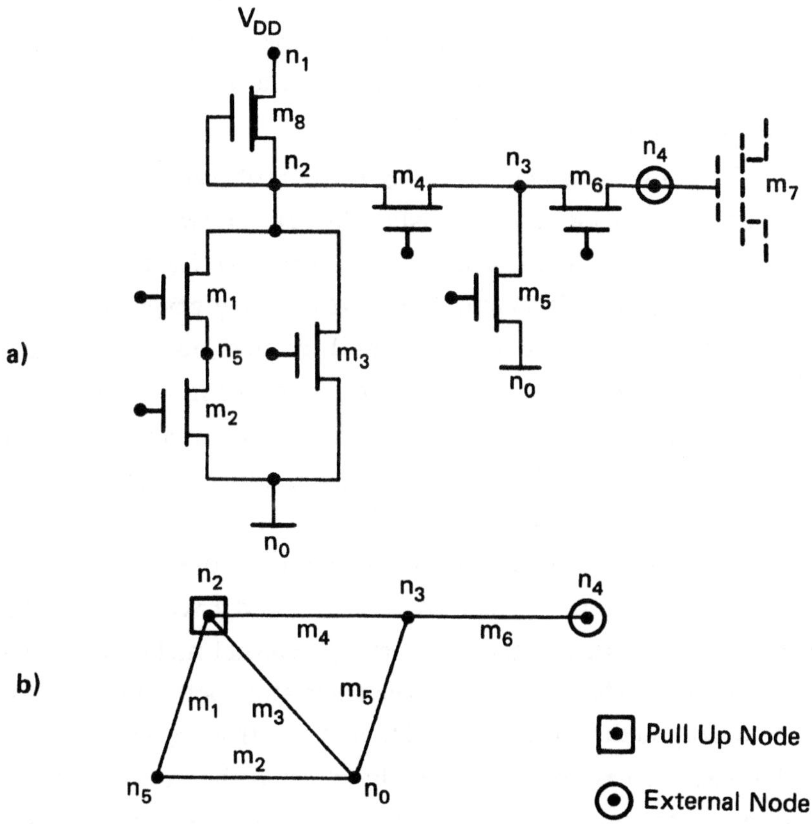

Figure 3.8(a) : An NMOS circuit with pseudo-external nodes
(b) : The graph representing the circuit in part (a)

signals at the pseudo-external nodes have to be computed in order to compute the signals at the external nodes of normal strength. Hence, such a node cannot be an internal node of an MFB. We, therefore, modify our definition of driver transistors to be the transistors in those paths between a pullup node and ground that do not contain an external or pseudo-external node. Thus, transistors m_4 and m_5 in the example in Figure 3.8(a) are not driver transistors.

The above modification is, however, still inadequate to be a necessary condition to be satisfied by driver transistors as is evident in the following example. Consider the NMOS network shown in Figure 3.9(a) and the corresponding graph in Figure 3.9(b). In this case we have two pullup nodes, namely, n_2 and n_4, and no external or pseudo-external nodes in the network. However, node n_3 cannot be considered an internal node in either of the two MFBs since its signal can be influenced by either of the two pullup nodes. Hence, the transistors m_4, m_5, and m_6 must be treated as pass transistors in this example. To include this case in our definition we would have to treat the other pullup nodes in the network as external nodes while we are trying to determine the driver transistors between a particular pullup node and ground. Thus, if we treat n_4 as an external node, then n_3 becomes pseudo-external; hence, we get m_1 and m_2 as the only driver transistors in the MFB corresponding to n_2. Similarly, if we treat n_2 as an external node we get m_3 as the only driver transistor in the MFB corresponding to n_4.

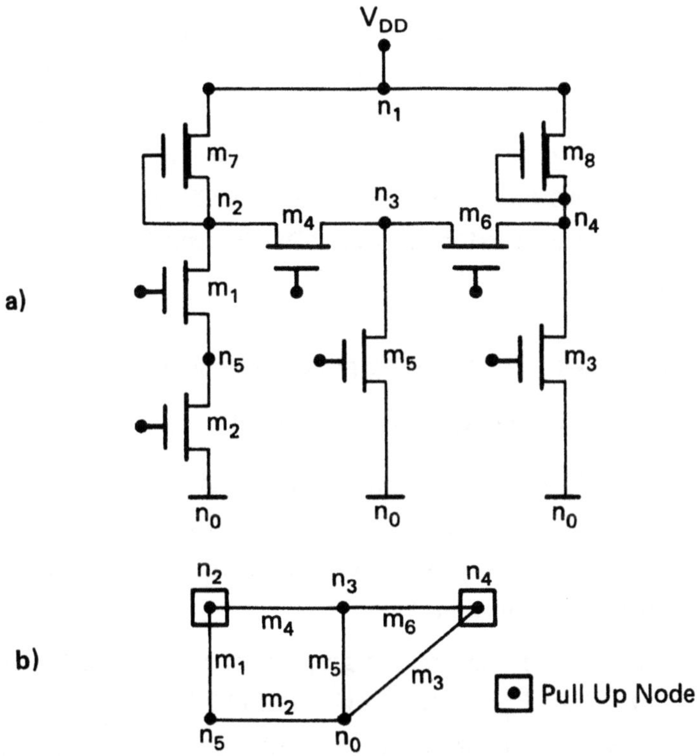

Figure 3.9(a) : An NMOS circuit with no external or pseudo-external nodes

(b) : The graph representing the circuit in part (a)

3.3.2 Formal Definitions

The purpose of the above discussion in Section 3.3.1 was mainly to help the reader form some kind of an intuitive idea on the difference between a driver and a pass transistor. Once again, we wish to emphasize that we are dealing with only NMOS networks. We now develop a completely precise definition of driver and pass transistors by formalizing the notion of external nodes in the NMOS network. Several modifications and extensions for CMOS circuits are discussed in Section 3.3.5.

Definition 3.4 : The set of *external* nodes is defined as

$$N_E = N_G \cup N_O \cup (N_I - \{GND\}) \tag{3.7}$$

the union of three sets, namely, the set of normal nodes which are gate nodes of enhancement transistors, the set of user-requested normal output nodes, and the set of input nodes without the ground node.

Let H be the undirected graph representing an NMOS network Ω. Let $E_{dep} = \{\phi^{-1}(m) : \mathrm{TRNTYP}(m) = \mathrm{n-dep}\}$ be the edges of H corresponding to the depletion transistors in Ω and let $\hat{H} = H - E_{dep}$. Let V_I, V_P, V_E denote the sets of input, pullup, and external vertices in H corresponding to the input, pullup, and external nodes in the network. Let $H_{IP} = ((\hat{H} \bullet V_I) \bullet V_P)$ be the graph obtained by splitting the input and pullup vertices in $\hat{H}$. The *strength* of a vertex v in H or $\hat{H}$ is the strength of the corresponding node $\theta(v)$ in the network Ω. Splitting a vertex retains the strength, i.e., the strength of the new vertices is the same as that of the original vertex before splitting. Also, splitting a vertex in a graph does not change the set of edges. Let C^H denote the subgraph of $\hat{H}$ induced by the edges in $E(C)$ for any component C of H_{IP}. Note that $E(H_{IP})=E(\hat{H})$; hence, C^H is well-defined. Consider a component C of H_{IP}. Then, clearly, C^H satisfies one and only one of the following conditions :

1 C^H contains at least one external vertex.
2(a) C^H contains no external vertices and no pullup vertices.
2(b) C^H contains no external vertices and exactly one pullup vertex.
2(c) C^H contains no external vertices and at least two pullup vertices.

Definition 3.5 : A component C of H_{IP} is said to be a *driver component* if C^H satisfies condition 2(b) given above.

Definition 3.6 : A component C of H_{IP} is said to be a *pass component* if C^H satisfies either condition 1, or 2(a), or 2(c) given above.

A component satisfying condition 2(a), i.e., having no external and no pullup vertices, is very rare since this represents a subnetwork containing only normal nodes, with the possibility of the ground node being included, while none of the normal nodes being gate nodes of enhancement devices or user-requested output nodes. Thus, this type of subnetwork neither interacts with other subnetworks nor is of any interest to the user. For the sake of completeness, however, we include this possibility also and label the component as a pass component.

The edges in a pass component are called *pass edges* while those in a driver component are called *driver edges*. It must be mentioned, once again, that splitting vertices in graphs does not alter the edge set of the original graph and so we have a partition of the edges of $\hat{H}$ into two sets, namely, the set of pass edges E_P and the set of driver edges E_D. We are now ready to define driver transistors and pass transistors in the NMOS network.

Definition 3.7 : An enhancement transistor **m** in the NMOS network $\mathbf{\Omega}$ is a *driver transistor* if $\phi^{-1}(\mathrm{m}) \in E_D$ and is a *pass transistor* if $\phi^{-1}(\mathrm{m}) \in E_P$.

We now form subgraphs with pass edges and driver edges and use these to define partitions of the NMOS network into special subnetworks. Let $H_I = \hat{H} \bullet V_I$ be the graph obtained by splitting all input vertices from $\hat{H}$. Let $H^1 = H_I - E_P$ be the graph obtained by removing all the pass edges from H_I and let $H^2 = H_I - E_D$ be the graph obtained by removing all the driver edges from H_I. Hence, H^1 contains only driver edges and H^2 contains only pass edges. The subgraph induced by the driver edges in a component of H^1 is called a *D-block* of $\hat{H}$ and the subgraph induced by the pass edges in a component of H^2 is called a *P-block* of $\hat{H}$. Once again, we make no distinction between edges in $\hat{H}$ and the graphs obtained by splitting some of its vertices since all these graphs have the same set of edges. We, thus, have partitioned the graph $\hat{H}$ into several edge-disjoint subgraphs H_i ; $i=1,2,\cdots,s$ where each H_i is either a D-block or a P-block. If H_i is a D-block then it must have a unique pullup vertex and no external vertices as a consequence of its definition. This fact and that in conventional NMOS designs a pullup node is connected to a unique depletion device allow us to make the following definition. The notion of an *induced subnetwork* is similar to that of induced subgraphs in a graph.

Definition 3.8 : A *multifunctional block* (MFB) is a subnetwork of Ω induced by the transistors corresponding to the edges of a D-block in $\hat{H}$ together with the depletion transistor connected to its pullup vertex (node). The pullup node is the *output node* of the MFB while the gate nodes of the driver transistors are its *input nodes*. The rest of the nodes, namely, the drain and source nodes of the driver transistors, apart from the pullup node and the ground node, are the *internal nodes* of the MFB.

Definition 3.9 : A *pass transistor block* (PTB) is a subnetwork of Ω induced by the transistors corresponding to the edges of a P-block in $\hat{H}$. Once again, the gate nodes of all the pass transistors are *input nodes* to the PTB. The rest of the nodes, namely, the drain and source nodes of the pass transistors, could either be input nodes, or output nodes, or both (sometimes called *ioputs* for both input and output), or none of the above depending upon the interaction of the PTB with the other blocks in the network. If a drain or source node of a pass transistor is of input strength, it is an input node to the PTB, if it is of pullup strength, it is an ioput (i.e., both input and output) node, and if it is a normal external node, it is strictly an output node of the PTB.

As an example, consider, once again, the circuit in Figure

3.9(a). The graph H_{IP} in this case, shown in Figure 3.10, has three components. The subgraph C_i^H in this example is the same as the component C_i itself, for each $i = 1,2,3$. The components, C_1 and C_3, clearly contain no external vertices and exactly one pullup vertex; hence, both are driver components according to Definition 3.5. The component C_2, however, contains no external vertices but has two pullup vertices and is, therefore, a pass component according to Definition 3.6. This circuit has two MFBs and one PTB. $\mathbf{MFB_1}$ consists of driver transistors $\mathbf{m_1}$, $\mathbf{m_2}$ and $\mathbf{MFB_2}$ has a single driver transistor $\mathbf{m_3}$, while the pass transistors $\mathbf{m_4}$, $\mathbf{m_5}$, and $\mathbf{m_6}$ constitute a **PTB**. A more detailed example is given in Section 3.3.4.

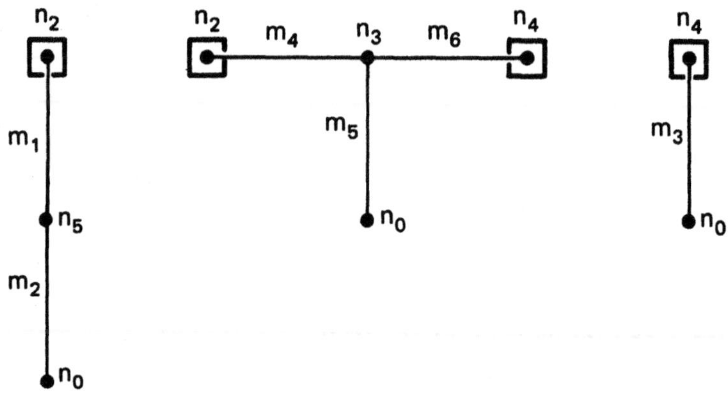

Figure 3.10 : The graph H_{IP} for H in Figure 3.9(b)

3.3.3 Partitioning Algorithm

We now describe the algorithm that partitions the set of enhancement transistors in an NMOS network $\Omega(N,M)$ into driver and pass transistors and then groups them to form MFBs and PTBs. Instead of dealing directly with $\Omega(N,M)$, we will be dealing with the graph H that represents the network. Once we have identified the D-blocks and P-blocks in H then, clearly, identifying the MFBs and PTBs is trivial. Hence, we shall mainly concentrate on the procedure PARTITION2 given below that partitions the graph H into several edge-disjoint subgraphs and labels each subgraph as either a D-block or a P-block.

Theorem 3.3 : The Algorithm 3.2 correctly partitions the edges of H into driver edges and pass edges and its time-complexity is $O(|V|+|E|)$ where V is the set of vertices and E is the set of edges in graph H.

Proof : The correctness of the algorithm can easily be verified since it partitions the edges of H directly according to Definitions 3.5 and 3.6.

In order to discuss the time-complexity, consider the procedures SPLIT and COMPONENT which are used several times in the above algorithm. If we can show that the time-complexity of each of these two procedures is $O(|V|+|E|)$, then we are done with the proof since the rest of the computations in PARTITION2 can easily be verified to be of linear time-complexity. From Theorem 3.2, we know that the splitting of vertices of V_I and V_P from H can be performed in $O(|E|)$ computation steps, since the splitting of a vertex from a graph

Algorithm 3.2 : Partitioning into Driver and Pass blocks

```
procedure PARTITION2 (H)
begin
        E_D ←Ø;
        E_P ←Ø;
        F^1 ←H ;
        for each v ∈ V_I do
                F^1 ←SPLIT(F^1,v );
        end for
        F^2 ←F^1;
        for each v ∈ V_P do
                F^2 ←SPLIT(F^2,v );
        end for
        Φ←COMPONENT(F^2);
        for each C_j ∈ Φ do
                n_E ←|V_E ∩ V(C_j^H)|;
                n_P ←|V_P ∩ V(C_j^H)|;
                if (n_P = 1 & n_E = 0) then
                        E_D ← E_D ∪ E(C_j );
                else
                        E_P ←E_P ∪ E(C_j );
                end if
        end for
        Φ_1←COMPONENT(F^1−E_P );
        for each H_i ∈ Φ_1 do
                BLK (H_i )←"D−block" ;
        end for
        Φ_2←COMPONENT(F^1−E_D );
        for each H_i ∈ Φ_2 do
                BLK (H_i )←"P−block" ;
        end for
        Φ ← Φ_1 ∪ Φ_2;
        return (Φ,BLK );
end
```

does not alter the edge sets. The procedure COMPONENT (F) returns the various components in the graph F and, once again, from Theorem 3.2 can be shown to have a time-complexity of $O(|V(F)| + |E(F)|)$. This procedure is used thrice in the above algorithm on graphs F^2, F^1-E_P, and F^1-E_D, respectively. However, $|V(F^2)|$, $|V(F^1-E^P)|$, and $|V(F^1-E^D)|$ are each $\leqslant |V(H)| + 2|E(H)|$, while $|E(F^2)|$, $|E(F^1-E_D)|$, and $|E(F^1-E_P)|$ are each $\leqslant |E(H)|$. Hence, we can conclude that the time-complexity of PARTITION2 (H) is $O(|V|+|E|)$. □

3.3.4 An NMOS Example

In this section we will consider the NMOS network shown in Figure 3.11 as an example to illustrate the partitioning into driver and pass transistors. This network consists of 17 nodes $\mathbf{N}=\{\mathbf{n}_0,\mathbf{n}_1,\ldots,\mathbf{n}_{16}\}$ and 20 transistors $\mathbf{M}=\{\mathbf{m}_1,\mathbf{m}_2,\ldots,\mathbf{m}_{20}\}$. The set $\mathbf{M}_{\mathbf{enh}}=\{\mathbf{m}_1,\mathbf{m}_2,\ldots,\mathbf{m}_{15}\}$ is the set of enhancement devices and $\mathbf{M}_{\mathbf{dep}}=\{\mathbf{m}_{16},\ldots,\mathbf{m}_{20}\}$ is the set of depletion devices. The set of nodes can be partitioned into three classes according to their strengths, namely, the nodes of input strength

$$\mathbf{N}_\mathbf{I}=\{\mathbf{n}_0,\mathbf{n}_1,\mathbf{n}_2,\mathbf{n}_3,\mathbf{n}_4,\mathbf{n}_5\},$$

the nodes of pullup strength

$$\mathbf{N}_\mathbf{P}=\{\mathbf{n}_6,\mathbf{n}_7,\mathbf{n}_8,\mathbf{n}_9,\mathbf{n}_{10}\},$$

and the nodes of normal strength

$$\mathbf{N}_\mathbf{N}=\{\mathbf{n}_{11},\mathbf{n}_{12},\mathbf{n}_{13},\mathbf{n}_{14},\mathbf{n}_{15},\mathbf{n}_{16}\}.$$

The node $\mathbf{n}_0$ is the ground node and $\mathbf{n}_1$ is the supply node to the network. The set of external nodes in this case is

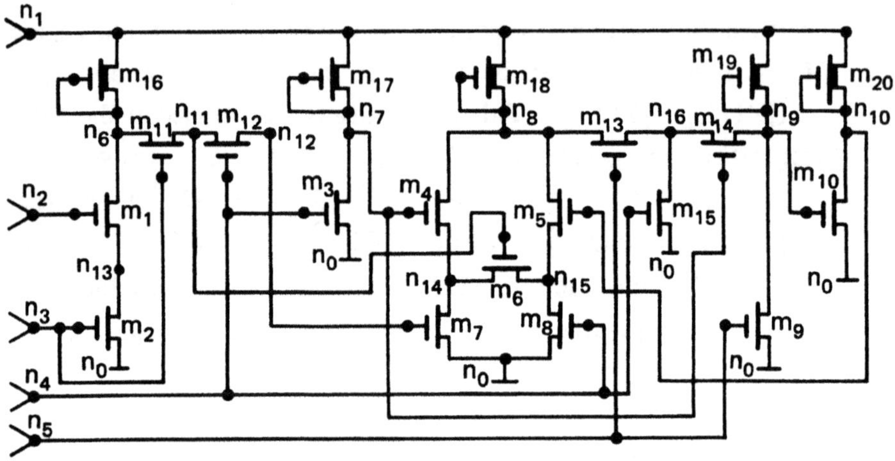

Figure 3.11 : An example of an NMOS network

$$N_E = \{n_2, n_3, n_4, n_5, n_{11}, n_{12}\}.$$

The graph H representing this network is shown in Figure 3.12. We only show the non-isolated vertices in the graph and only those edges corresponding to the enhancement transistors (i.e., we are dealing with $\hat{H}$ as defined in Section 3.3.2). Also, we refer to the vertices and edges of the graph as nodes and transistors in the network, respectively, for the sake of convenience, i.e., in this case the bijections θ and ϕ used in Definition 3.1 are both identity mappings. The graph H_I obtained by splitting the nodes of input strength from H is shown in Figure 3.13 and the graph H_{IP} is shown in Figure 3.14(a). The graph H_{IP} has 7 components. The subgraphs of H

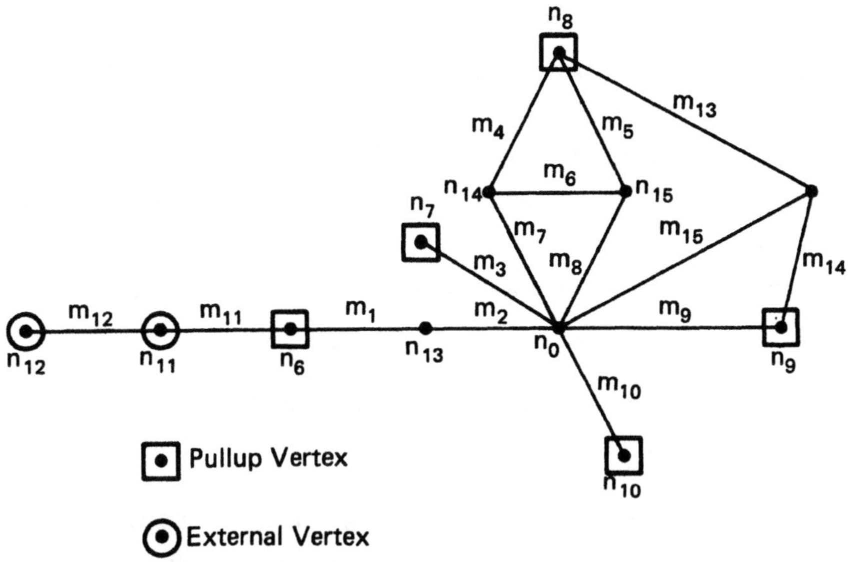

Figure 3.12 : The graph H for the network in Figure 3.11

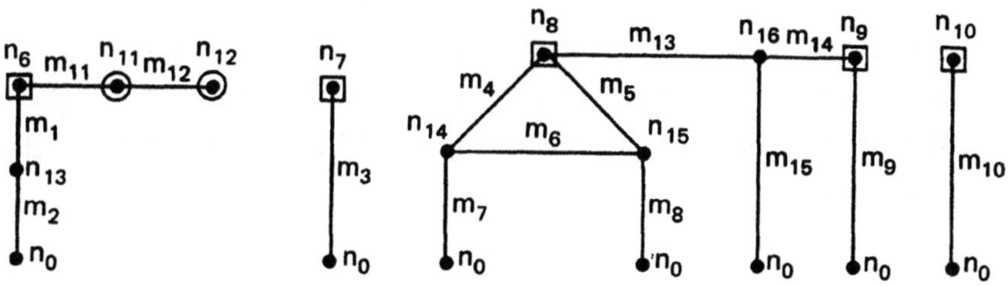

Figure 3.13 : The graph H_I for H in Figure 3.12

induced by the edges in each of these components are shown in Figure 3.14(b). Among these subgraphs, the subgraph C_2^H has 2 external nodes while C_5^H has 2 pullup nodes, and so the corresponding components C_2 and C_5 are declared as pass-components. The rest of the components can easily be verified to be driver-components. Thus, the set of driver transistors is

$$M_D=\{m_1,m_2,m_3,m_4,m_5,m_6,m_7,m_8,m_9,m_{10}\}$$

and the set of pass transistors in the network is

$$M_P=\{m_{11},m_{12},m_{13},m_{14},m_{15}\}$$

It can easily be verified that the subgraphs C_2^H and C_5^H are the P-blocks of H and the rest of the subgraphs in Figure 3.14(b) are the D-blocks of H. Thus, the transistors in the network can be now partitioned into 7 blocks, 2 of which are PTBs and the remaining 5 are MFBs. Table 3.4 contains a listing of the transistors in each block along with the set of its input and output nodes. In the case of an MFB the first transistor in its list is a depletion load device.

In addition to the 7 blocks given in Table 3.4, the network also has 5 SRCs which are listed in Table 3.5 along with the node of input strength in each of them. We have, thus, partitioned the network shown in Figure 3.11 into 5 SRCs, 5 MFBs and 2 PTBs.

3.3.5 Modifications for CMOS Circuits

We will now consider a few modifications that are needed in the partitioning scheme, to include CMOS networks as well.

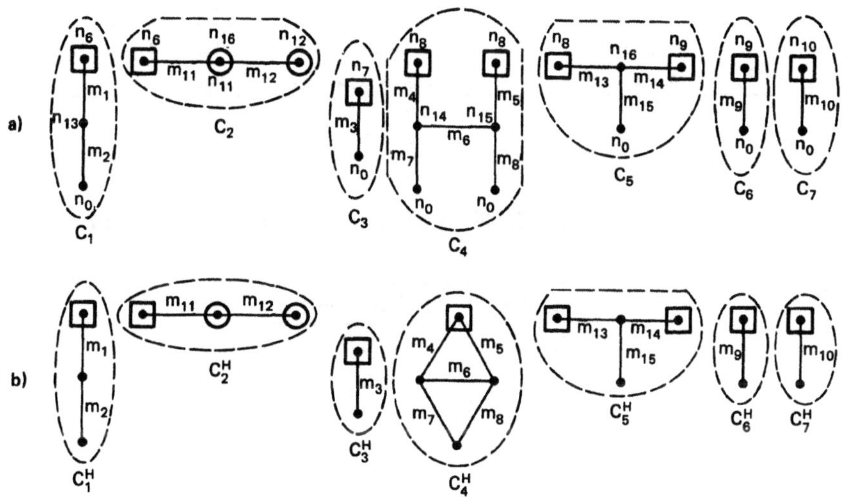

Figure 3.14(a) : The graph H_{IP} for H in Figure 3.12
(b) : The edge-induced subgraphs of H

Table 3.4: A list of blocks of the network in Figure 3.11

Block	Transistors	Input Nodes	Output Nodes
MFB_1	m_{16},m_1,m_2	n_2,n_3	n_6
MFB_2	m_{17},m_3	n_4	n_7
MFB_3	$m_{18},m_4,m_5,m_6,m_7,m_8$	$n_4,n_7,n_{10},n_{11},n_{12}$	n_8
MFB_4	m_{19},m_9	n_5	n_9
MFB_5	m_{20},m_{10}	n_9	n_{10}
PTB_1	m_{11},m_{12}	n_3,n_4,n_6	n_6,n_{11},n_{12}
PTB_2	m_{13},m_{14},m_{15}	n_4,n_5,n_7,n_8,n_9	n_8,n_9

Table 3.5: A list of SRCs of network in Figure 3.11

Block	Output Node
SRC_1	n_1
SRC_2	n_2
SRC_3	n_3
SRC_4	n_4
SRC_5	n_5

In such networks the graph used to represent the network would consist of two types of edges, namely, n-type and p-type edges, corresponding to n-channel and p-channel transistors respectively. In conventional CMOS circuits, pass transistors are usually implemented using an n-channel and a p-channel device having common drain and source nodes. The edges corresponding to these transistors can easily be detected and removed from the graph. Once this is done, a pullup node can be identified as a node adjacent to both n-type and p-type edges in the resulting graph. One can then use the scheme described earlier in this section to complete the partitioning. An MFB in CMOS would consist of a network of n-channel devices between the pullup node and ground and a dual network of p-channel devices between the pullup node and V_{DD}. A PTB would also consist of both n-channel and p-channel pass transistors.

3.4 ORDERING OF PARTITIONED BLOCKS

Let $\Omega(N,M,\Sigma)$ be the MOS network that has been partitioned into several blocks. The network will be processed by processing each of its blocks in a certain order. A block is said to be processed, if given the waveforms at the input nodes to the block, the waveforms at its output nodes are obtained. Thus, in order to process a block, the waveforms at its input nodes must be known. Hence, we must process the blocks in a certain order so that this condition is always satisfied (whenever possible). In this section we will show when such an ordering exists, and if so, how does one obtain it.

Definition 3.10 : For each node $n_i \in N$ in the network, let $FOUT(n_i)$ denote the *fanout-list* for the node which is the set of blocks in Σ having n_i as an input node, and let $FIN(n_i)$ denote its *fanin-list* which is the set of blocks with n_i as an output node. Thus,

$$\mathrm{FOUT}(n_i) = \{\Omega_j : n_i \in \mathrm{INP}(\Omega_j)\} \tag{3.8}$$

and

$$\mathrm{FIN}(n_i) = \{\Omega_j : n_i \in \mathrm{OUT}(\Omega_j)\}. \tag{3.9}$$

Let $(\Omega_j, \Omega_k, n_i)$ denote an ordered triple in $\Sigma \times \Sigma \times N$. The ordered triple $(\Omega_j, \Omega_k, n_i)$ is said to be an *I/O-triple* if $\Omega_j \in \mathrm{FIN}(n_i)$ and $\Omega_k \in \mathrm{FOUT}(n_i)$. For applications where it becomes necessary to distinguish between driver and pass transistors within a partitioned block, the I/O-triples are further classified as adjacent and nonadjacent triples. The interested reader may refer to [78] for details. Using the I/O-triples in the network, we will now introduce the notion of a good ordering in which the blocks of a network could be processed.

Definition 3.11 : A *sequential ordering* **R** on the blocks of a partitioned network $\Omega(N,M,\Sigma)$ is a 1-1 function $R : \Sigma \rightarrow \{1,2, \cdots ,s\}$ where $s = |\Sigma|$. The sequential ordering **R** is said to be a *good-ordering* for the network if $R(\Omega_j) < R(\Omega_k)$ for every I/O-triple $(\Omega_j, \Omega_k, n_i)$ in the network.

A good ordering, as defined above, is clearly a desirable ordering for processing the blocks in a network, since in this case, whenever a block is scheduled for processing, all the blocks in the fanin-lists of each of its input nodes have been previously processed, thus, providing input signals to the this block. A good ordering, however, may not exist for some networks. As an example, consider a block Ω_k in a network having its output node n_o connected back to one of its inputs. In this case, the network is said to have feedback, and the definition of a good ordering would be violated by the I/O-triple $(\Omega_k, \Omega_k, n_0)$ for any sequential ordering. Hence, there is no good ordering for such a network. In the remaining part of this chapter we will show that a good ordering exists only for networks not having any kind of feedback, and proceed to handle the case of a network with feedback. The latter is important since most of the networks designed in present day MOS technology do have feedback in some form or another, for example, flip-flops, ring oscillators, and most clocked sequential circuits in general. To this end, we will use the notion of a directed graph derived from a partitioned-network. But, first, we review some basic concepts on directed graphs from Bondy and Murty [55], for the sake of readers not very familiar with the subject.

3.4.1 Directed Graphs

A *directed graph* G, often abbreviated as a *digraph*, is formally defined as an ordered triple $(V(G), A(G), \psi_G)$ consisting of a nonempty set $V(G)$ of *vertices*, a set, $A(G)$, disjoint from $V(G)$, of *arcs*, and an *incidence function* ψ_G that associates with each arc of G an ordered pair of (not necessarily distinct) vertices of G. If a is an arc and v and w are vertices such that $\psi_G(a) = (v, w)$, then a is said to *join* v to w; v is the *tail* of a, and w is its *head* and the arc is usually referred to as simply (v, w). A digraph G' is a *subdigraph* of G if $V(G') \subseteq V(G), A(G') \subseteq A(G)$ and the incidence function $\psi_{G'}$ is the restriction of ψ_G to $A(G')$. With each digraph G we can associate an undirected graph H on the same vertex set ;

corresponding to each arc of G there is an edge of H with the same ends. The graph H is said to be the *underlying graph* of G. The terminology and notation for subdigraphs are similar to those used for subgraphs. Just as with graphs, digraphs also have a simple pictorial representation. A digraph is represented by a diagram of its underlying graph together with arrows on its edges, with each arrow pointing towards the head of the corresponding arc. Figure 3.15(a) shows a digraph G and its underlying graph H is shown in Figure 3.15(b).

A *directed walk* in G is a finite nonempty sequence $W = (v_0, a_1, v_1, \cdots, a_k, v_k)$, whose terms alternate between vertices and arcs, such that, for each $i=1,2,\cdots,k$ the arc a_i has head v_{i-1} and tail v_i. Directed *paths* and *cycles* are similarly defined. The vertex v_0 is called the *origin* of the directed path while v_k is its *terminus*, and the rest of the vertices are called internal vertices. The integer k denotes the length of the directed path. Once again, the integer k denotes the length of the directed cycle. A directed cycle of length k is referred to as a *k-cycle*. If there exists an arc a in G such that $\psi_G(a) = (v, v)$, then a is a *loop* in G, and v, a, v is an example of a 1-cycle in G. As with paths and cycles in undirected graphs, we will also refer to the subdigraphs induced by the arcs in a directed path or cycle as a directed path or cycle. Furthermore, for convenience, we will drop the term "directed" and refer to directed paths and directed cycles simply as paths and cycles.

A path in G with origin v and terminus w is called a (v,w)-path. If there is a (v,w)-path in G then the vertex w is

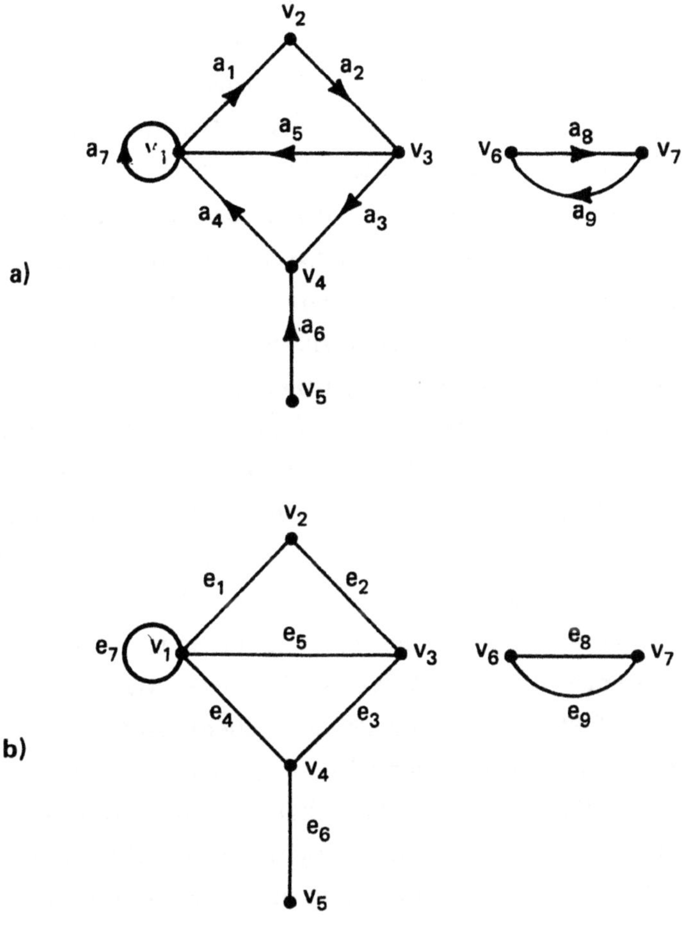

Figure 3.15(a) : A digraph $G(V, A)$
(b) : The underlying graph $H(V, E)$

said to be *reachable* from v in G. This, however, does not imply that v is also reachable from w. Two vertices v and w are said to be *strongly-connected* in G, denoted by $v \sim w$, if each is reachable from the other. Clearly, $\sim$ is an equivalence relation on $V(G)$ and it partitions $V(G)$ into nonempty subsets $V_1, V_2, \ldots, V_\mu$, such that if $v \in V_i$ and w is strongly connected to v in G, then w must also be $\in V_i$. The subdigraphs $G[V_1], G[V_2], \ldots, G[V_\mu]$ induced by the partition are called the strongly-connected components of G. Note, by definition, a vertex v in G is always strongly to itself, i.e., $v \sim v$ since one can always choose a directed path of length 0 and reach v from itself and *vice versa*. Thus, $G[V_i]$ is a *trivial* strongly-connected component if it contains only one vertex, i.e., $|V_i| = 1$. It can be easily shown that if $G[V_i]$ is a non-trivial strongly-connected component of G, i.e., $|V_i| \geqslant 2$, then it must necessarily contain a k-cycle with $k \geqslant 2$. Thus, presence of non-trivial strongly connected components in a digraph implies the presence of directed cycles. We use $\mu(G)$ to denote the number of strongly-connected components in G. We say that G itself is strongly-connected if $\mu(G) = 1$. Figure 3.16(a) shows a digraph which has three strongly-connected components as shown in Figure 3.16(b). Hence, the digraph is not strongly-connected, while its underlying undirected graph is connected, since it has only one component. This clearly illustrates the difference between strongly-connectedness in digraphs and connectedness in undirected graphs.

The *in-degree* $d_G^-(v)$ of a vertex v in G is the number of arcs having v as their head vertex. Similarly, the *out-degree*

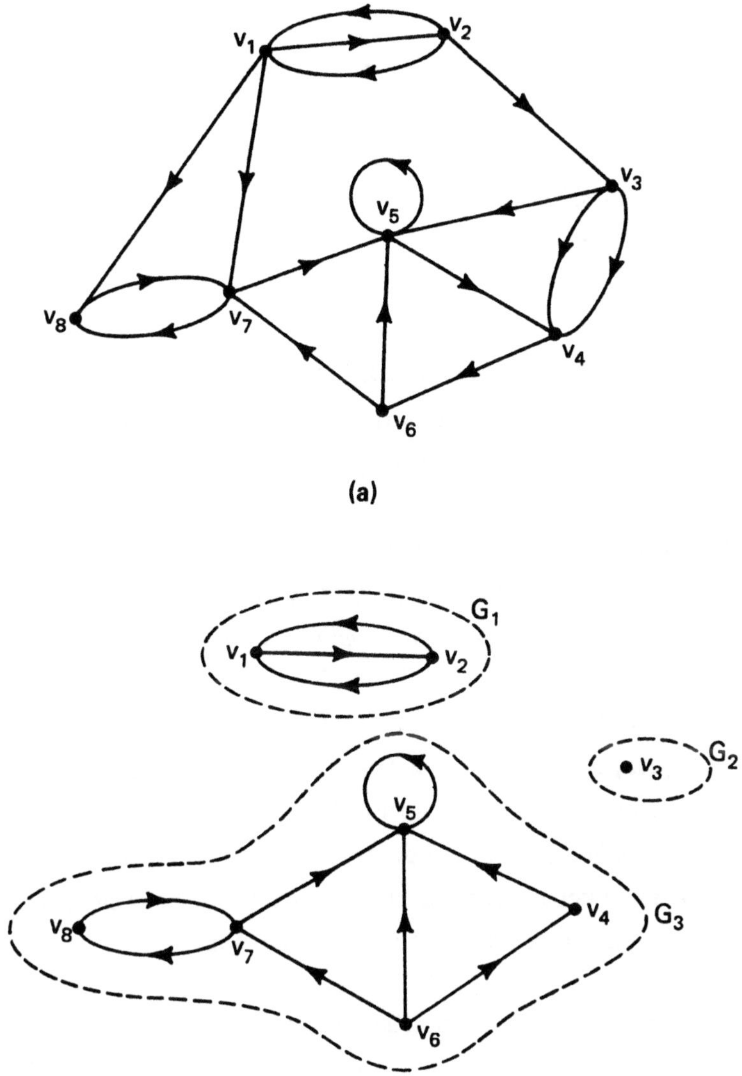

Figure 3.16(a): A digraph G
(b): The three strongly-connected components of G

$d_G^+(v)$ of a vertex v is the number of arcs having v as their tail vertex. Just as with undirected graphs, we shall use the symbols $\nu(G)$ and $\epsilon(G)$ to denote the number of vertices and arcs in G. We shall also drop the letter G from most of the notations whenever possible.

3.4.2 Presence of Feedback and Its Detection

Let $\mathbf{\Omega}(\mathrm{N,M,\Sigma})$ be a partitioned network. Let Υ denote the set of I/O-triples of the network.

Definition 3.12 : A directed graph $G(V, A, \psi_G)$ is said to be *derived* from an NMOS partitioned-network $\mathbf{\Omega}(\mathrm{N,M,\Sigma})$ if there exist bijections $\theta{:}\Sigma \rightarrow V$ and $\phi{:}\Upsilon \rightarrow A$ such that the triple $v = (\Omega_j, \Omega_k, n_i) \in \Upsilon$ is an I/O-triple in the network if and only if $\psi_G(\phi(v)) = (\theta(\Omega_j), \theta(\Omega_k))$. Thus, for every I/O-triple $v = (\Omega_j, \Omega_k, n_i)$ of the partitioned network, there is an arc $a = \phi(v)$ in the derived digraph G with tail vertex $\theta(\Omega_j)$ and head vertex $\theta(\Omega_k)$. The digraph G is said to be *acyclic* if it has no directed cycles. Just as with blocks in a network, we have sequential orderings on vertices of a digraph.

Definition 3.13 : A sequential ordering R on the vertices of a digraph G is said to be a *topological-ordering* if for every arc a with tail v and head w the strict inequality $R(v) < R(w)$ is satisfied.

Theorem 3.4 : If $\mathbf{\Omega}$ is a partitioned NMOS network and G is its derived digraph, then the following three conditions are equivalent :

(1) There is a good ordering on the blocks of $\mathbf{\Omega}$,

(2) G is acyclic, and

(3) There exists a topological ordering on the vertices of G.

Proof : We shall first show that (1) => (2). Suppose **R** is a good ordering on the set of blocks $\mathbf{\Sigma}$ of the network. We will show that the derived digraph cannot contain a directed cycle. Suppose G has a directed k-cycle. If $k=1$, then there is a loop a with both ends at some vertex v. By Definition 3.12, there exists an I/O-triple $(\mathbf{\Omega}_j,\mathbf{\Omega}_j,\mathbf{n}_i)=\phi^{-1}(a)$, where $\mathbf{\Omega}_j=\theta^{-1}(v)$ in the network. This I/O-triple would, clearly, violate Definition 3.11 for the good ordering **R**. If $k>1$ then let v denote the vertex in the k-cycle C with the corresponding block $\mathbf{\Omega}_j=\theta^{-1}(v)$ ordered first by **R** among blocks corresponding to the other vertices in the cycle, i.e., $\mathbf{R}(\theta^{-1}(v)) \leqslant \mathbf{R}(\theta^{-1}(w))$ for all $w \in C$. Since $k>1$ there is an arc a from w to v in the cycle (and therefore in G) with $w \neq v$. But this would mean that there is an I/O-triple $(\mathbf{\Omega}_i,\mathbf{\Omega}_j,\mathbf{n}_i)$ in the network where $\mathbf{\Omega}_i=\theta^{-1}(w)$, thus leading to $\mathbf{R}(\mathbf{\Omega}_i)<\mathbf{R}(\mathbf{\Omega}_j)$, i.e., $\mathbf{R}(\theta^{-1}(w))<\mathbf{R}(\theta^{-1}(v))$, which contradicts the above choice of the vertex v. Hence, the proof is by contradiction.

The fact that (2) => (3) is a well-known result on digraphs and can be found in most standard textbooks on graph theory, such as [55]. Hence, we will only outline this part of the proof. Suppose G is an acyclic digraph. Then there must be a vertex of in-degree 0 in G. If not, consider the longest directed path in G. If the first vertex of this path does not have in-degree 0, then either G has a cycle or a longer path, which is a contradiction. Hence, G has a vertex, say v, whose in-degree is 0. The rest of the proof that G has a topological ordering is by induction on the number of vertices of G. The basis for induction is clearly satisfied for all digraphs containing only one vertex. Now suppose that all acyclic digraphs on less than ν vertices have a topological ordering. Let G have ν vertices. Then $G-v$ has no cycles and has $\nu-1$ vertices, and so must have a topological ordering, say R'. Now let R be an ordering of G such that $R(v)=1$ and $R(w)=R'(w)+1$ for all other vertices $w \neq v$ in G. Then clearly, R is a topological ordering for G.

The fact that (3) => (1) follows, trivially, from the definitions of good orderings of Σ, topological orderings of vertices in G and the fact that G is derived from Ω. □

We now introduce the concept of feedback in a partitioned network.

Definition 3.14 : A partitioned MOS network Ω is said to have *feedback* among its blocks if its derived digraph G has directed cycles. Thus, Ω is *feedback-free* if G is acyclic and is *internal feedback-free* if G has no directed loops. A block $\Omega_j \in \Sigma$ is said to have *internal feedback* if the corresponding vertex $\theta(\Omega_j)$ in G is incident with a directed loop.

It is clear that this definition of feedback in the networks conforms to the standard notion of feedback in circuits. We state the following result without proof, below, since it clearly follows from Theorem 3.4 and the definition of feedback-free networks.

Theorem 3.5 : A partitioned-network $\Omega(N,M,\Sigma)$ has a good ordering on its partitioned blocks if and only if it is feedback-free.

A good ordering of the blocks in a feedback-free network can easily be obtained by first placing the vertices of the derived digraph (which in this case will be acyclic, by definition) in a topological order and then placing the corresponding blocks of the network in the same order. If, however, the network has feedback (which is the more common case in the present-day MOS designs), the derived digraph contains directed cycles; hence, no topological (good) ordering exists on its vertices (blocks). In this case, therefore, one must detect the blocks in the network that are within feedback-loops, treat these as special blocks and and place the rest of the blocks in a "good" ordering. We formalize these ideas below.

Definition 3.15 : If V_i is a set of vertices in a strongly-connected component of G, then the corresponding set $\Sigma_i = \{\theta^{-1}(v) : v \in V_i\}$ of blocks in Σ is defined to be a *strongly-connected component* (SCC) of the network. Thus, we have a partition $\Sigma_1, \Sigma_2, \ldots, \Sigma_\mu$ of the blocks in Σ.

Let $V_1, V_2, \ldots, V_\mu$ denote the partition of the vertex set of the digraph G into strongly-connected components. We define the *condensation* of G to be a digraph $\tilde{G}$ consisting of vertices $w_1, w_2, \ldots, w_\mu$ with an arc having head w_i and tail w_j if and only if $i \neq j$ and there is an arc in G with head $x \in V_i$ and tail $y \in V_j$. Consider the digraph G shown in Figure 3.16(a). Its condensation $\tilde{G}$, shown in Figure 3.17, is clearly acyclic. We will show that, for any digraph G, its condensation $\tilde{G}$ is acyclic; hence, from Theorem 3.4, it has a topological ordering, which corresponds to an ordering of the SCCs of Σ. To this end we need the following intermediate result.

Lemma : If C is a directed cycle in a digraph G, then all its vertices must be within a strongly-connected component of G.

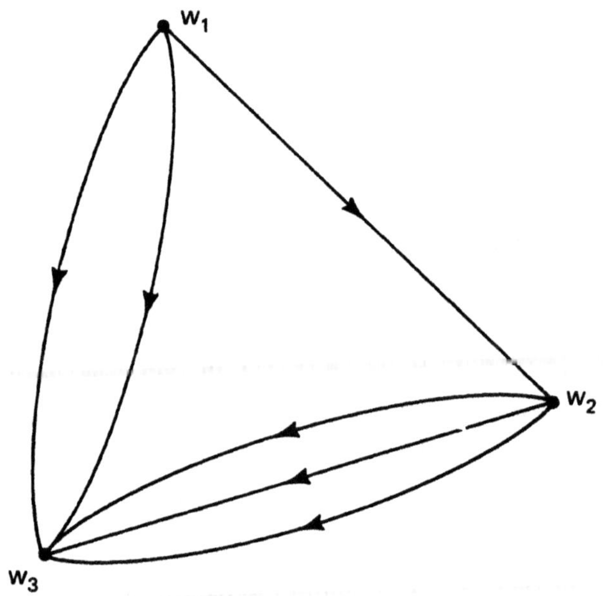

Figure 3.17 : The condensation of the digraph G in Figure 3.16(a)

Proof : (See [55]). Consider any two vertices, say, x and y in $V(C)$. Since C is a cycle, there is a directed path from x to y and also a return path from y to x in C. But C is a subdigraph of G; hence, x is reachable from y and y is reachable from x in G. Therefore, by definition, x and y must be in the same strongly-connected component. □

Theorem 3.6 : The condensation $\tilde{G}$ of any digraph G must be acyclic.

Proof : (See [55,58]). By definition, $\tilde{G}$ has no directed loops. Therefore, $\tilde{G}$ has no 1-cycles. If $\tilde{C}$ is a k-cycle in $\tilde{G}$ with $k > 1$. then the vertices of G in the set $\bigcup_{w_j \in \tilde{C}} V_j$ must belong to a directed cycle; hence, from the above lemma, these vertices must all be in one strongly-connected component, which is a contradiction. □

Our strategy to schedule the blocks of Σ for processing is to start by detecting the strongly-connected components in the derived digraph G. We then obtain the condensation of G and proceed to find a topological ordering in $\tilde{G}$. This then corresponds to some ordering on the SCCs of Σ. The processing of the network Ω then begins by processing the SCC ordered first, followed by the one ordered second and so on. An SCC is said to be *simple* if it contains only one block of Σ and that block has no internal feedback. A simple SCC is processed by algorithms described in Chapter 4. If an SCC is not simple then the blocks within it are processed using special techniques described in Chapter 5.

Algorithm 3.3 presented below, well-known as Tarjan's algorithm [54], partitions the vertex set of any digraph into its strongly-connected components. A vertex w is an *in-neighbor* of the vertex v in G if (w, v) is an arc of G and is an *out-neighbor* of v if (v, w) is an arc of G. We use $Adj_G^-(v)$ and $Adj_G^+(v)$ to denote the sets of in-neighbors and out-neighbors of the vertex v in G. In Tarjan's algorithm two integers $k[v]$ and $L[v]$ are computed for each vertex v in the digraph G, known as depth-first number and lowpoint [58] respectively.

A digraph G is said to be a *rooted-digraph* if it contains a vertex, say *root*, such that all vertices in G are reachable from *root*. In the case of the derived digraph G we make it a rooted-digraph by inserting a new vertex called *root* and directing arcs from this new vertex to every vertex of in-degree 0 in the original G. In the original derived digraph G every vertex corresponding to an SRC block in the circuit must indeed have in-degree 0 and so the above notion is well defined. Further, if there is a vertex that is not reachable from the new vertex *root*, then it is also not reachable from any of the vertices corresponding to the SRC blocks in the network. This means that the input signals would never propagate to such blocks in the network and so they need not be simulated. Hence, we are only interested in simulating those blocks in the circuit that correspond to vertices that are reachable from the vertex *root* in the above new digraph. We will still refer to the modified derived digraph as G itself and will assume that it is a rooted-digraph.

The fact that the Algorithm 3.3 terminates for finite digraphs and does so with linear time-complexity and,

Algorithm 3.3 : Detecting SCCs

```
procedure SCC_DETECT (G)
begin
        i ←1;
        for each v ∈ V do
                MARK [v ]← "new";
        end for
        μ←0;
        initialize STACK to empty;
        v ←root ;
        DFS (v );
end
procedure DFS (v )
begin
        MARK [v ]← "old";
        k [v ]←i ;
        i ←i +1;
        L [v ]←k [v ];
        push v on STACK;
        for each vertex w ∈ Adj_G^+(v ) do
                if MARK [w ] = "new" then
                        DFS (w );
                        L [v ]←MIN (L [v ],L [w ]);
                else if k [w ] < k [v ] and w ∈ STACK then
                        L [v ]←MIN (k [w ],L [v ]);
                end if
        end for
        if L [v ] = k [v ] and v ≠root then
                μ←μ+1;
                V_μ←Ø;
                repeat
                        pop x from STACK;
                        V_μ←V_μ ∪ {x };
                until x = v ;
        end if
end
```

furthermore, correctly partitions the vertices of the digraph into strongly-connected components follows from the theorem stated below without proof. Its proof can be found in several books on graph algorithms such as [54,56,58].

Theorem 3.7 : The procedure SCC_DETECT (G) partitions the vertices of V into its strongly-connected components correctly with time-complexity of $O(\max(|V|,|A|))$.

We now describe Algorithm 3.4 that creates a new digraph $\tilde{G}$ which is the condensation of the digraph G. We will use two procedures CREATE (x) and ADD_ARC (x,y) to create vertices and add arcs in the data structure that represents $\tilde{G}$. This algorithm has time-complexity $O(\nu+\epsilon)$, where $\nu = |V(G)|$ and $\epsilon = |A(G)|$.

Algorithm 3.4 : Condensation of a digraph

```
procedure CONDENSE (G)
begin
        V(G̃)←∅;
        A(G̃)←∅;
        for i←1 until μ do
                CREATE (w_i);
        end for
        for each arc (x,y)∈A(G) do
                i←SCCOMP(x);
                j←SCCOMP(y);
                if i≠j then
                        ADD_ARC (w_i,w_j);
                end if
        end for
        return G̃;
end
```

We finally present Algorithm 3.5 that produces a topological ordering on the vertices of the digraph $\tilde{G}$ which is known to be acyclic from Theorem 3.6; hence, from Theorem 3.4, there exists such an ordering. This algorithm uses a QUEUE to store some vertices. One could also use a STACK instead which would result in a different ordering. The topological ordering R on $\tilde{G}$ provides us with an ordering **ORD** on the set of SCCs in the network.

Algorithm 3.5 : Topological ordering [56]

```
procedure TOP_ORDER (G̃ )
begin
        k ←1;
        for each vertex w_i ∈Ṽ do
                I [w_i]←d⁻(w_i);
                if d⁻(w_i) = 0 then
                        push w_i into QUEUE;
                end if
        end for
        while QUEUE is not empty do
                pop vertex w_j from QUEUE;
                R [w_j ]←k ;
                k ←k +1;
                for each vertex w_k ∈Adj⁺(w_j ) do
                        I [w_k ]←I [w_k ]−1;
                        if I [w_k ] = 0 then
                                push w_k into QUEUE;
                        end if
                end for
        end while
        return R ;
end
```

3.4.3 An Example to Illustrate Ordering

Consider once again the NMOS example of Section 3.3.4 shown in Figure 3.11. This network was partitioned into 12 blocks, namely, 5 SRCs, 5 MFBs, and 2 PTBs. The fanin and fanout lists for each node in the network are given in Table 3.6 for those nodes for which both lists are non-empty. From this table, we see that, node $\mathbf{n_6}$ is both an input node as well as an output of $\mathbf{PTB_1}$ since it a pullup node. This would generate the I/O-triple ($\mathbf{PTB_1}$,$\mathbf{PTB_1}$,$\mathbf{n_6}$) which would prohibit any ordering from being a good ordering of the network. Such a triple is called a nonadjacent I/O-triple [78] and must be excluded while

Table 3.6: Fanin and fanout lists of nodes of Figure 3.11

Node	Fanin List	Fanout List
n_2	SRC_2	MFB_1
n_3	SRC_3	MFB_1,PTB_1
n_4	SRC_4	MFB_2,MFB_3,PTB_1,PTB_2
n_5	SRC_5	MFB_4,PTB_2
n_6	MFB_1,PTB_1	PTB_1
n_7	MFB_2	MFB_3,PTB_2
n_8	MFB_3,PTB_2	PTB_2
n_9	MFB_4,PTB_2	PTB_2,MFB_5
n_{10}	MFB_5	MFB_3
n_{11}	PTB_1	MFB_3
n_{12}	PTB_1	MFB_3

constructing the derived graph; otherwise, a directed self loop would indicate feedback. Similarly the I/O-triples (PTB_2,PTB_2,n_7) and (PTB_2,PTB_2,n_8) are also excluded from the original list. The remaining 19 triples are used to construct the derived digraph G as shown in Figure 3.18. For precise definitions of and details on adjacent and nonadjacent I/O-triples the reader is referred to [78].

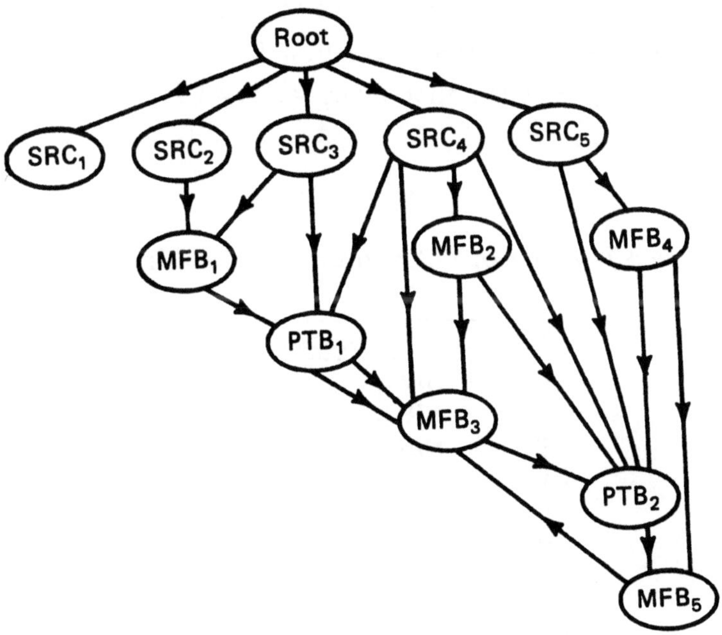

Figure 3.18 : The derived graph G

This digraph contains 10 vertices and 19 arcs. We also include a vertex *root* and join it to the 5 SRC vertices as shown. Using Algorithm 3.3 on this digraph we get 10 strongly connected components which are listed in Table 3.7. Thus, $\mathbf{\Sigma_{10}}$ is the only SCC that is not simple. Since G has no self-loops, the network has no internal-feedback. Note that had we considered the three I/O-triples we excluded in constructing the derived digraph, we would get self-loops. The network, however, has an SCC that is not simple; hence, it has feedback among $\mathbf{MFB_3}$, $\mathbf{PTB_2}$, and $\mathbf{MFB_5}$. The condensation digraph $\tilde{G}$ is shown in Figure 3.19. From Algorithm 3.5 on $\tilde{G}$ we get a topological ordering R such that $R(w_i)=i$; $i=1,2,\ldots,10$. This

Table 3.7: A list of SCCs in network of Figure 3.11

SCC	Blocks
Σ_1	**SRC_1**
Σ_2	**SRC_2**
Σ_3	**SRC_3**
Σ_4	**SRC_4**
Σ_5	**SRC_5**
Σ_6	**MFB_1**
Σ_7	**MFB_2**
Σ_8	**MFB_4**
Σ_9	**PTB_1**
Σ_{10}	**MFB_3,MFB_5,PTB_2**

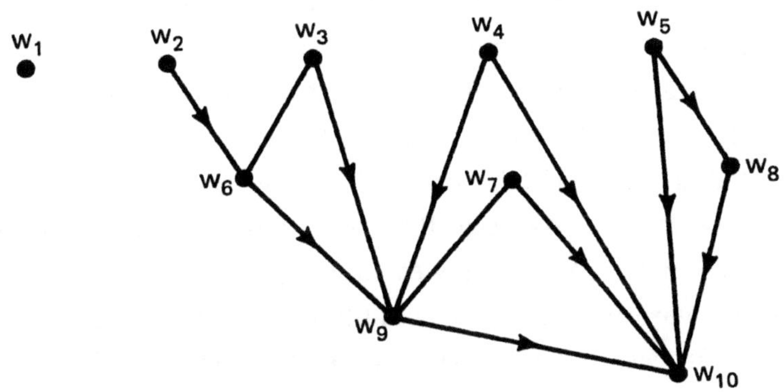

Figure 3.19 : The condensation graph $\tilde{G}$

induces an ordering **ORD** on the SCCs of the network such that $\mathbf{ORD}(\Sigma_i) = \mathbf{i}$; i=1,2, . . . , 10, in this case.

3.5 CONCLUSIONS

In this chapter we began by representing an MOS network $\Omega(N,M)$ as a set of nodes N interconnected by a set of MOS devices M. We then partitioned the entire network into several transistor-disjoint channel-connected subnetworks or blocks. In Section 3.3, we further partitioned the set of enhancement transistors into driver transistors and pass transistors. Following this, a set of driver transistors were grouped together to form an MFB while pass transistors were grouped together to form a PTB. Another type of block, called SRC, was introduced to model the input voltage sources connected to the input nodes of the network. The partitioned network was represented as $\Omega(N,M,\Sigma)$, where Σ is the set of partitioned blocks. We then introduced the concept of feedback among blocks in the network and showed that a good ordering for processing the various blocks is possible only for feedback-free networks. In case the network has feedback, the set of blocks were partitioned into its strongly-connected components (SCCs). Finally, we came up with an ordering of the SCCs for processing. The partitioning and the ordering of the blocks have both been shown to take computation time that is linear in the number of circuit nodes and number of devices.

CHAPTER 4

SWITCH-LEVEL TIMING SIMULATION

Let $\mathbf{\Omega}(N,M,\mathbf{\Sigma})$ be a partitioned MOS network in which the set of blocks $\mathbf{\Sigma}$ has been further partitioned into its strongly-connected components (SCC) $\mathbf{\Sigma}_1,\mathbf{\Sigma}_2,\ldots,\mathbf{\Sigma}_\mu$. Let **ORD** denote the ordering in which the SCCs have been scheduled for processing as described in Section 3.4. If an SCC is simple, i.e., it consists of exactly one block with no internal feedback, then it is simulated at the switch level by algorithms described in this chapter. By simulating or processing a block, we mean, obtaining the digital waveforms at the output(s) of the block given those at the inputs to the block over the entire time-interval of interest. In case the SCC is not simple, a special event-driven windowing technique, to be described in Chapter 5, is used to simulate the various blocks within the SCC. This special technique partitions the entire time-interval into several windows and uses the algorithms described in this chapter to simulate only the active blocks within each window.

4.1 OVERVIEW

The core of a switch-level timing simulator is the method in which a delayed output waveform is determined, given the input waveforms to a subcircuit block. The goal is to determine the shape of output waveforms, without expending an excessive amount of computer resources in finding the exact voltages at every time-point. Many tradeoffs must be made between computational speed and memory constraints on the one hand and accuracy on the other in order to obtain the desired results. The most important tradeoffs are between waveform representation and the delay method.

In the context of switch-level timing simulation, as defined in Section 2.5, the user is only interested in obtaining p-state digital equivalents of the analog waveforms at various nodes in the circuit over a certain time-interval. Clearly, the larger the number of states, the better is the level of detail provided, and thus, the more useful the information is to the user. However, the complexity of digital simulation dramatically increases with the number of states **p**, particularly in the context of generating accurate timing estimates. The choice of **p=2** must be rejected outright, since in this case only binary (i.e., 0 or 1) waveforms are produced. Binary waveforms contain no information on the slopes of the corresponding analog waveforms, the presence of glitches, or other information which is often useful to a designer when evaluating the performance of a circuit. In our model, therefore, we use three states (i.e., **p**=3) to describe the values of digital signals, which seems to be a fair compromise between the level of detail and the generation of accurate SLT estimates.

There are several tradeoffs with regard to the delay calculation method. Some methods may require a great deal of presimulation to build tables which will be accessed during simulation [31,32,34,35,38]. The number and size of the tables depend on their complexity, waveform representation, and the number of subcircuit "primitives." Other methods have greater flexibility in that no presimulation is required, although more simulation time is spent performing calculations [33]. The accuracy of the method depends on the inclusion of several factors. These include input waveform slopes, transistor model parameters, device capacitances, parasitics, overlapping inputs, waveform voltage levels, and transistor configurations within subcircuits. All of these issues must be addressed in order to obtain the accuracy which is required by designers.

In order to obtain the desired simulation speed, all subcircuit blocks will be processed by mapping them into equivalent "primitive" subcircuit blocks [32]. The assumption is that by carefully analyzing the primitive blocks, complex blocks can be simulated to within a desired accuracy using the equivalent model of an appropriate primitive block. Since most blocks in digital circuits are small, the primitive blocks are usually chosen to have the same configuration as those that are frequently used in digital design. Difficulties may arise when analyzing subcircuits with pass transistors and unusual transistor configurations. The greater the number of primitives, the greater is the complexity of the modeling method. A subcircuit block must be analyzed to determine which primitive will best model it. The block must then be mapped into that primitive in order to find the delayed waveforms.

There are two approaches to finding the delayed output waveform of a block. The first approach first determines a zero-delay output waveform, and then applies delays to the output transitions. The zero-delay waveform is determined using switch-level simulation methods. An important aspect of this switch-level simulation is the determination of where transitions will occur. If an output node has connections to both V_{DD} and GND, the size of the devices must be compared to determine the final output state. The waveform may then be delayed by applying delays determined from tables or empirical equations found through presimulation. After the delays are applied, the waveform must be checked for consistency (overlapping output transitions must be filtered) [32].

There are three methods of finding the delays of a zero-delay waveform. The first is a tabular method, which works well for NMOS circuits. The depletion load transistor in NMOS circuits serves as a method to give nearly all rising waveforms the same slope. Tables are built through presimulation to give delays for each primitive for wide combinations of device sizes, input slopes, and output loading [31,32,34,35,38]. During simulation the correct delay is found using table lookup. The second method also uses tables, but in a different form. Each transistor is modeled as a linear resistor, converting a subcircuit block into an RC network. The delay through the RC network is just the RC time-constant. The delay tables are used to map the transistors to resistors, where resistance is a function of the slope of the waveform at the gate node. This method is adopted by *timing analyzers* [46-53]. A third method is based on constructing delay equations based on the information in the tables [37]. Unfortunately this type of empirical equation is difficult to determine and cannot remain accurate over a wide range of parameters.

The above three methods of delay computation are much more difficult to implement for CMOS circuits, which may have a wide range of connections to both V_{DD} and GND. This requires much more complex tables, which would require at least an order of magnitude more presimulations than for NMOS circuits.

The second approach to finding delayed output waveforms, which is followed in this book, calculates the voltage changes on the output nodes of the block during simulation [33] without having to perform switch-level simulation first. This is quite different from the first approach in that no circuit presimulation is performed and the equations do not directly yield delay values. These equations are derived from the transistor dc characteristic equations for a simple inverter, as shown in Figure 4.1. This approach has a great deal of flexibility in terms of process parameters, waveform voltages, and overlapping inputs. Unlike tabular methods, a change in process parameters has no effect on simulation speed or preprocessing time, as will be seen later.

A frequently overlooked, but important, issue is the determination of the dc output voltage, V_{net}, of a subcircuit block. If the devices in a subcircuit block have inappropriate sizes, there may not be enough current available to charge (or discharge) the output nodes. When there are connections to both V_{DD} and GND, V_{net} must be determined in order to decide if any output node transition will occur. If a subcircuit block contains several NMOS pass transistors, the output voltage will not rise completely to V_{DD}. There may be a drop of several volts. This may introduce errors if the waveform is assumed to switch completely. The effect of the errors will arise as incorrect delay values in subsequent blocks with the erroneous waveform as an input.

Another issue in the switch-level timing simulation of a subcircuit block is the method in which the block is reduced to that of a primitive. The reduction may be performed using an

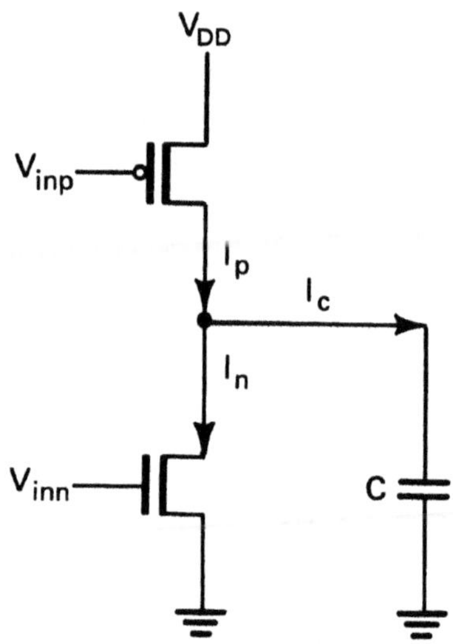

Figure 4.1: Basic inverter structure for equations

internal node elimination technique, including simple series-parallel reductions [32]. A difficulty with this method is the lack of consideration of internal node capacitances. If the internal node capacitances are not small, there may be significant errors in the delay calculation. A second approach uses linear RC network processing to reduce the circuit to a simple RC time-constant. The difficulty with this approach is in the calculation of the delay for such a reduction. This is due to the fact that transistors cannot be accurately modeled as linear resistors.

4.2 WAVEFORM REPRESENTATION

One of the first issues to be considered in timing simulation is the representation of signal waveforms. In circuit simulation it is necessary to store the voltage of every node at every time-point. In switch-level timing simulation the general shape of the waveform is of primary concern rather than individual voltages at a large number of time-points. In other words, it is only necessary to calculate and store only those time-points when the response crosses some threshold values.

In logic or unit-delay simulation the slope of a switching waveform is not included; only one switching time-point is computed as the time when the response switches from low to high or high to low. To obtain more accurate results, switch-level timing simulation takes into consideration the slope of an input transition in order to calculate delays at the output of a subcircuit block. Thus, a minimum of two switching threshold points is required. The use of more than two threshold points is, however, not cost-effective; it requires more overhead and increases the computation time without any significant gain in accuracy. Since two threshold points are adequate to approximate waveform slopes during transitions, we have chosen two threshold points.

Let $[t_0, t_f]$ denote the time-interval in which the network $\Omega(N,M)$ is to be simulated. At each time-instant, the signal at a node in the network is assumed to have a *ternary logic value*, i.e., a 0, I, or 1, while this value might change with time. Such a signal is called a *ternary logic signal.* A node $n \in N$ is associated with a *ternary digital waveform*, $X_n(t)$ which is related to its analog counterpart $V_n(t)$ as follows :

$$X_n(t) = \begin{cases} 0 \text{ if } 0.0 \leqslant V_n(t) \leqslant V_{TL} \\ I \text{ if } V_{TL} < V_n(t) < V_{TH} \\ 1 \text{ if } V_{TH} \leqslant V_n(t) \leqslant V_{DD} \end{cases} \tag{4.1}$$

where V_{TL} and V_{TH} are two thresholds chosen such that $0.0 < V_{TL} < V_{TH} < V_{DD}$. Here, I is an *intermediate* state between the steady low and high states 0 and 1 used to represent signals in transition. A *transition* in a ternary signal is defined as a change in the ternary value of the signal taking place at a certain time-instant. Thus, to completely specify a transition, we need to specify both the type of transition and the time at which it occurs. There are four types of transitions possible. These are $0 \to I$, $I \to 1$, $1 \to I$, and $I \to 0$. We can now represent a *transition* α by $(t, x \to y)$ where $x \to y$ is one of the four allowable transition types and t denotes the time of transition. Furthermore, x is the *initial* value of α and y is its *final* value.

Let $S = \alpha_1, \alpha_2, \ldots, \alpha_p$ be a *sequence* of transitions where $\alpha_j = (t_j, x_j \rightarrow y_j)$ for each $j=1,2,\ldots,p$. The sequence S is said to be *chronological* if $t_1 < t_2 < \cdots < t_p$. A chronological sequence is said to be *compatible* if, in addition, $y_j = x_{j+1}$ for each $1 \leqslant j \leqslant p-1$. In a compatible sequence, therefore, the final value of every transition in the sequence is equal to the initial value of the succeeding transition.

Ternary waveforms can now be efficiently stored as a compatible sequence of transitions. Figure 4.2(a) shows an analog waveform, V_{in}, which is represented by the sequence

$$(t_1, 1 \rightarrow I), (t_2, I \rightarrow 0), (t_3, 0 \rightarrow I), (t_4, I \rightarrow 1). \qquad (4.2)$$

This sequence is then interpreted as shown in Figure 4.2(b).

Consider a chronological and compatible sequence of transitions $S = \alpha_1, \alpha_2, \ldots, \alpha_p$ representing a ternary digital waveform, where $\alpha_j = (t_j, x_j \rightarrow y_j)$. Suppose $t_0 = t_{start}$ and $t_{p+1} = t_{stop}$. The ternary waveform partitions the entire time-interval into *sub-intervals* (t_j, t_{j+1}) for each $j=0,1,2,\ldots,p$ such that during each of these intervals the ternary digital signal has a fixed logic value, i.e., a logic 0, I, or 1. We now introduce the notion of the *state* of a ternary signal during the interval (t_j, t_{j+1}). A ternary signal can occupy one of *five* different states, namely, **LO**, **MID**, **HI**, **SWLO**, or **SWHI**, indicating low, middle, high, switching low, or switching high, respectively. During the first interval (t_0, t_1) the state of the ternary signal is defined by

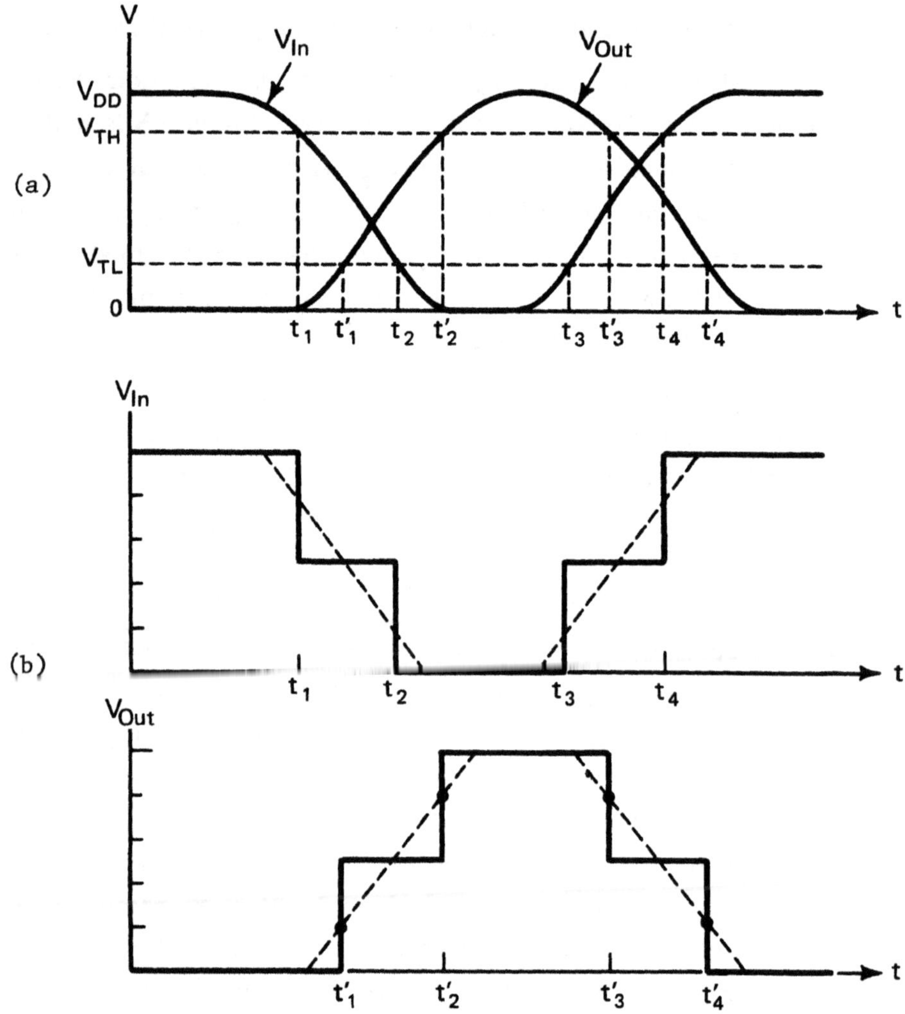

Figure 4.2. Ternary logic delay definition
(a) Analog representation
(b) Digital representation

$$\text{state} = \begin{cases} \text{LO if } x_1 = 0 \\ \text{MID if } x_1 = I \\ \text{HI if } x_1 = 1 \end{cases} \tag{4.3a}$$

Similarly, during the last interval (t_p, t_{p+1}) the state is defined by

$$\text{state} = \begin{cases} \text{LO if } y_p = 0 \\ \text{MID if } y_p = I \\ \text{HI if } y_p = 1 \end{cases} \tag{4.3b}$$

For each j=1,2,...,p−1 the state during interval (t_j, t_{j+1}) is defined by

$$\text{state} = \begin{cases} \text{LO if } y_j = 0 \\ \text{HI if } y_j = 1 \\ \text{MID if } y_j = I \text{ and } x_j = y_{j+1} \\ \text{SWHI if } x_j = 0\text{ , } y_j = I\text{ , and } y_{j+1} = 1 \\ \text{SWLO if } x_j = 1\text{ , } y_j = I\text{ , and } y_{j+1} = 0 \end{cases} \tag{4.3c}$$

A signal in the **MID** state is assumed to correspond to an analog voltage of $(V_{TH} + V_{TL})/2$. Consider once again V_{in} shown in Figure 4.2(b). A signal having this waveform is in state **HI** during (t_{start}, t_1), in state **SWLO** during (t_1, t_2), in state **LO** during (t_2, t_3), in state **SWHI** during (t_3, t_4), and finally in state **HI** during (t_4, t_{stop}).

The ternary logic value of a signal at a time-point and the state of the signal during an interval are two different notions. A signal can occupy one of five different states during an interval but can have only one of three logic values at any time-point during that interval. A signal occupying the **LO** (**HI**) state during some interval has a ternary value of 0(1) at each time-point in that interval. On the other hand, a signal occupying either the **MID**, **SWLO**, or **SWHI** state during some interval has, in each case, a ternary value I at each time-point within the interval.

In the remaining sections of this chapter we will be dealing only with the state of a node-signal and not with its ternary value. However, in some cases, we will be interested in storing the exact analog voltage reached when signals are in either the **LO** or **HI** states. To this end, we modify the previously used representation of signals to include voltage limits on the switching waveforms. An additional parameter is added to each transition with a final value of 0 or 1 to signify the voltage reached after each transition. The sequence for the ternary waveform in Figure 4.2(b) would then be

$$(t_1, 1 \rightarrow I),(t_2, I \rightarrow 0, 0.0 \text{ volt}),(t_3, 0 \rightarrow I),(t_4, I \rightarrow 1, 5.0 \text{ volts}). \quad (4.4)$$

The selection of the logic threshold levels is important in retaining accuracy and preventing unrealistic negative delay values. If the threshold values are relatively close to each other, the accuracy of the slope represented by the threshold crossing times is reduced. This could also lead to negative delay values when a slow input waveform causes a fast output transition. In addition to being inaccurate, negative delay values introduce many difficulties when handling circuits with feedback as will be seen in Chapter 5, and thus, should be avoided. In CMOS circuits, if V_{TL} is set near the device threshold voltage for an NMOS driver transistor and V_{TH} is set near the value of V_{DD} plus the PMOS driver transistor device threshold, the delays should always be positive and the thresholds are far enough apart to retain good accuracy. The delays are positive in this case because the output will not change until a transistor is turned ON or OFF, which will not occur until a threshold voltage is crossed by an input signal. In NMOS circuits V_{TH} should be set just under the highest voltage a normal external node voltage reaches.

4.3 SIMULATION ALGORITHM

The simulation algorithm is outlined here in order to aid in the understanding of the remainder of this chapter. Let Ω_q be a subcircuit block with a set of transistors M_q to be simulated in the time-interval $[t_{start}, t_{stop}]$. Let

$$NDS=\{DRAIN(m_i),\ SOURCE(m_i) : m_i \in M_q\}$$

be the set of drain and source nodes of the transistors in the block and let $NG=\{GATE(m_i) : m_i \in M_q\}$ be the set of gate nodes. Let INP and OUT be the set of input nodes and output nodes of the block as defined by Equations (3.5) and (3.6) respectively. Let N_I denote the set of nodes in NDS of input strength and $INTERNAL = NDS - N_I - OUT$ denote the *internal nodes* in the block.

Consider the set $\{t_1,t_2,\ldots,t_f\}$ of transition times of the ternary signals at the gate nodes arranged in an ascending order $t_1<t_2<\cdots<t_f$, where $t_1=t_{start}$ and $t_f=t_{stop}$. These time-points divide the time-interval of simulation into several *phases* such that during each phase $\phi_j=(t_j,t_{j+1})$ the signal at each gate node in **NG** is at a fixed state (see Section 4.2 for definition of state). The time t_j is the *initial time* and the time t_{j+1} is the *final time of phase* ϕ_j. The procedure PHASE divides the simulation interval $[t_{start},t_{stop}]$ into phases and stores them in a set Θ.

The task of switch-level timing simulation is to compute the sequences of transitions at the nodes in **OUT**, given the ternary signals at nodes in **INP**. This is done in phases. First the block in the first phase ϕ_1 is simulated, followed by the next phase and so on, updating the sequences of transitions for the output nodes in each phase.

In each phase ϕ_j a transistor m_i in the block can occupy one of five states depending on the nature of the digital waveform at its gate node. Let n_i=GATE(m_i) be the gate node of m_i having state $s_{i,j}$ in phase ϕ_j. The state of a transistor m_i during phase ϕ_j is then defined by the following function:

```
function STATE (m_i,φ_j)
begin
        if (TRNTYP(m_i) = n—enh) then
                if (s_i,j = HI) then
                        STATE ←ON;
                else if (s_i,j = LO) then
                        STATE ←OFF;
```

```
                else if (s_i,j = MID) then
                        STATE ←INT;
                else if (s_i,j = SWHI) then
                        STATE ←SWON;
                else if (s_i,j = SWLO) then
                        STATE ←SWOFF;
                end if
        else if (TRNTYP(m_i) = p—enh) then
                if (s_i,j=HI) then
                        STATE ←OFF;
                else if (s_i,j = LO) then
                        STATE ←ON;
                else if (s_i,j = MID) then
                        STATE ←INT;
                else if (s_i,j = SWHI) then
                        STATE ←SWOFF;
                else if (s_i,j = SWLO) then
                        STATE ←SWON;
                end if
        else if (TRNTYP(m_i) = n—dep) then
                STATE ←ON;
        end if
end
```

In the above definition of state of a transistor we have introduced two new states **SWON** and **SWOFF** to represent transistors that are switching in addition to the three states **ON**, **OFF**, and **INT**. A transistor in state **INT** is assumed to have a gate voltage of $(V_{TL}+V_{TH})/2$.

The simulation of a phase ϕ_j begins by constructing an undirected graph **H** with vertex set **V=NDS** corresponding to the drain and source nodes of the transistors in the block. For each transistor $m_i \in M_q$, an edge is inserted between **DRAIN**(m_i) and **SOURCE**(m_i) if and only if STATE(m_i,ϕ_j) is **ON**, **SWON**,

INT, or **SWOFF**. There is no edge corresponding to a transistor in the **OFF** state. The edge corresponding to transistor m_i is assigned a state equal to that of the transistor and a resistance equal to $\frac{2L}{K_pW}$. The procedure BUILDGRAPH outlined below constructs the required graph.

```
procedure BUILDGRAPH(Ω_q, φ_j, H)
begin
    V←{DRAIN(m_i),SOURCE(m_i) : m_i ∈ M_q} ;
    E←∅;
    for each transistor m_i ∈ M_q do
        S←STATE(m_i,φ_j);
        if S is not OFF then
            u←DRAIN(m_i);
            v←SOURCE(m_i);
            ADD_EDGE (e , u , v , E),
            RES(e)← 2L/(K_p W);
            state(e)←S;
        end if
    end for
end
```

The graph is then reduced by eliminating the nodes with negligible capacitance. This procedure is described in Section 4.6. Then, for each output node the state and time-constant are found for the paths from both V_{DD} and **GND**. The time-constants are found by the PARSEGRAPH procedure, which is outlined in Section 4.6. Although not specified, the PARSEGRAPH routine also finds the resistance of the paths from the voltage sources to the output node. These resistances, R_{VDD} and R_{GND}, are used to include the effects of parasitics in the time-constant. The resistances are also used by the DCOUT procedure, which finds the net voltage, V_{net}, the output node is being driven towards. The procedure DCOUT is described in Section 4.5. To find V_{net} all switching transistors are assumed to switch completely. A **SWON** transistor is treated as an **ON** transistor and a **SWOFF** transistor is treated as **OFF**. Based on this assumption, V_{net} is found using the methods discussed in Section 4.5. If V_{net} is equal to the initial value of V_{out} in this phase, then there is no change in the output node voltage, and calculations are not needed. If V_{net} is not equal to the initial value of V_{out}, then calculations of output node voltage changes must be performed.

In this case, the effects of parasitics are first included in the time-constants. The time-constants, τ^*_{VDD} and τ^*_{GND}, are then used in the DELAY routine to find the voltage changes on the output nodes.

The SIMULATE procedure will now be outlined. This procedure should be used as a reference throughout the remainder of this chapter.

```
procedure SIMULATE(Ω_q, t_start, t_end)
begin
      PHASE (Ω_q , t_start , t_stop , Θ) ;
      for each phase φ_j ∈ Θ do
            t_width ← duration of phase φ_j ;
            BUILDGRAPH (Ω_q, φ_j, H) ;
            REDUCEGRAPH (H) ;
            for each output node n ∈ OUT(Ω_t) do
                  V_out ← final V_out of n in previous phase;
                  PARSEGRAPH (H, n, V_DD, τ_VDD, S_p) ;
                  PARSEGRAPH (H, n, GND, τ_GND, S_n) ;
                  DCOUT (Ω_q , φ_j , n , V_net) ;
                  if (V_net ≠ V_out) then
                        include parasitics in τ_VDD and τ_GND;

                        DELAY(V_net, V_out, t_width, τ*_VDD,

                                    τ*_GND, S_p, S_n);
                  end if
            end for
      end for
end
```

4.4 DERIVING INVERTER VOLTAGE EQUATIONS

In this section the equations that are used to estimate the delay values in subcircuit blocks are derived from the transistor current equations of a single inverter. Thus, there is only one primitive. All subcircuit block output nodes are charged or discharged by the connections to V_{DD} and GND. These connections to V_{DD} and GND are reduced to an equivalent inverter circuit so that the voltage equations derived for a single inverter can be applied. The mapping of complicated subcircuit blocks into an equivalent single inverter case will be discussed later in Section 4.6.

Figure 4.1 shows the configuration of a CMOS inverter used to derive the equations. Although referred to as an inverter, there is no assumption that the transistor inputs are tied together. The gate of the p-channel device has a signal V_{inp} while the gate of the n-channel device has a signal V_{inn}. The output node capacitance C is the sum of the internal transistor capacitances and the output line and fanout capacitance. This capacitance value gives accurate results, provided the line capacitance is much greater than the sum of the internal device capacitances. The output capacitance is charged (and discharged) by the current I_C, which is the difference between I_P and I_N. More formally,

$$C\frac{dV_{out}}{dt} = I_C(t, V_{out}) \tag{4.5}$$

or

$$\frac{dV_{out}}{dt} = \frac{I_P(t, V_{out})}{C} - \frac{I_N(t, V_{out})}{C}. \tag{4.6}$$

By substituting the correct current equations for I_P and I_N from the transistor characteristic equations into (4.6) and integrating, equations giving the change in output voltage as a function of time are found. These nonlinear equations of output voltage change are applied during simulation by inserting the proper parameters, which reflect subcircuit configuration, model parameters, and loading.

The calculations of voltage changes take place in phases of time. During each phase, the output node has some specific connection to V_{DD} and GND. The phases serve as a method for dividing time into sections, where one type of calculation of the output voltage change is made in each phase. A new phase is defined by the change of state of a transistor, such as a second overlapping input switching ON. Every time an input changes, the subcircuit block must be reparsed to find the new connections from V_{DD} and GND to the output nodes, if such connections exist. The objective is to find the time-points at which the output voltage crosses the logic threshold voltages. There are several other time-points which must be found. These are the times at which the transistors change regions of operation. These must be determined in order to perform correct calculations.

In order to derive equations based on the dc characteristics of a transistor, many assumptions must be made. The first assumption is that the device threshold, V_t, is fixed. Including the body and drain effects would complicate the equations beyond practical usage. Many other model parameters must be ignored, but there are some parameters which could be included in the equations for more accuracy. These include temperature, length and width variations on the device threshold, length and width reduction factors, source and drain resistance, channel length modulation, mobility modulation, velocity saturation, and device transconductance (K_p) variations between the linear and saturated regions. Many of these factors would apply only to short channel devices. For simplicity, the equations derived here will not include these factors. The form and usage of the equations are derived based on the following dc characteristic equations:

$$I_D = \frac{K_p W}{2L}(V_{GS}-V_t)^2 \qquad \text{for } |V_{GS}| \geqslant |V_t| \qquad (4.7)$$

$$\text{and } |V_{DS}| \geqslant |V_{GS}-V_t|$$

$$I_D = \frac{K_p W}{2L}\left[2(V_{GS}-V_t)V_{DS}-V_{DS}^2\right] \text{ for } |V_{GS}| \geqslant |V_t| \qquad (4.8)$$

$$\text{and } |V_{DS}| \leqslant |V_{GS}-V_t|$$

$$I_D = 0 \qquad \text{for } |V_{GS}| < |V_t| \qquad (4.9)$$

For the n-type driver transistor $I_D = I_N$, $V_{GS} = V_{inn}$, and $V_{DS} = V_{out}$. For the p-type transistor $I_D = I_{SD} = I_P$, $V_{GS} = V_{inp} - V_{DD}$, and $V_{DS} = V_{out} - V_{DD}$. For the n-type load transistor $V_{GS} = 0$, and $V_{DS} = V_{DD} - V_{out}$. The equations are derived using the CMOS gate, but only a simple substitution is required for NMOS gates.

An important observation is the fact that the current equations for the inverter gate are essentially symmetric. The equations which are used to perform the calculations of a rising output voltage are also used on a falling output voltage. In addition to switching the n-type and p-type parameters, the uses of V_{TL} must be replaced by $V_{DD} - V_{TH}$ and V_{TH} must be replaced by $V_{DD} - V_{TL}$. Thus, the equations found here are based only on a rising output voltage. The equations contain the terms V_{DD}, V_{TH}, and V_{TL}. Node voltages are not assumed to make a full swing between V_{DD} and GND. A value of V_G will appear in the equations rather than V_{DD}. This allows correct calculations when an input waveform does not make a full swing, which allows greater accuracy and flexibility in the simulator.

The equations are derived using certain notations. If the change in output voltage is to be found for a given time-step, t_{step}, the initial output voltage is V_{outi} and the final voltage is V_{outf}. The amount of time required for the output voltage to change from V_{outi} to a desired output voltage of V_{outd} is t_{req}. The phase width, or the length of time for which the state of the transistor inputs of the inverter remains the same, is t_{width}. Thus, after t_{width} a new set of equation parameters will be required. Frequently a time t_{req} is found for the output to reach V_{outd}, such as $V_{outd} = V_{TH}$. If t_{req} is greater than t_{width}, then the final output voltage V_{outf} at the end of t_{width} time is calculated.

The transistor current equations (4.7) and (4.8) are analyzed in two forms. The first form is for those phases of time in which the input of one or both of the transistors is switching. In many cases the current function is in a form which can be explicitly integrated. However, the case of a switching waveform on the input of a transistor in the linear region cannot be explicitly integrated. The current in this case is usually primarily dependent on time rather than the output voltage, V_{out}. Thus, in order to integrate equation (4.6) in this case, the currents are converted to polynomials in time. When

V_{out} would appear in the equation, an estimate of $V_{out,est}$ is made as

$$V_{out,est} = V_{outi} + \mathbf{B}\ t. \tag{4.10}$$

V_{outi} is the current output voltage and **B** is an estimate of the slope of the output waveform. This estimate is used only for a short period of time, and in the cases where it is used, the primary factor in the current equations is the switching gate voltage. Thus, the error introduced by an inaccurate estimate over a small interval of time is minor.

A switching waveform at the transistor gate is represented as

$$V_{GS} = V_{GSi} + \mathbf{S}\ t, \tag{4.11}$$

where V_{GSi} is the voltage at the beginning of the analysis interval and **S** is the slope of the switching input. If the input of a transistor is rising, **S** is positive. The value of **S** is easily obtained from the waveform representation as

$$\mathbf{S} = \frac{V_{TL} - V_{TH}}{t_2 - t_1} \tag{4.12}$$

for the falling waveform in (4.2).

The second form of the analysis is used when the gate voltages of both transistors are fixed. In this case the current is a function of the output voltage only. This permits equation (4.6) to be solved in two ways. First, it is solved to give the amount of time, t_{req}, required to reach a desired output voltage V_{outd}. Second, if the phase of time in which the inputs remain constant, t_{width}, is less than the required time, t_{req}, the equations are solved to give the final output voltage V_{outf} at the end of that phase.

To simplify notation, the following substitution is used:

$$Y = \frac{K_p W}{2L} \tag{4.13}$$

This parameter, Y, is referred to as the conductance of the transistor, which is the inverse of the transistor resistance as defined in Section 4.3. This represents the transconductance term of the nonlinear transistor currents.

4.4.1 Equations for Switching Inputs

When the input to a transistor is switching, the current equation is a nonlinear function of both time t and the output voltage V_{out}. Since a function of this form is difficult to integrate explicitly, the output voltage is estimated as a linear function of t, as in (4.10). Thus, it is necessary to control the estimate of B to keep the error within reasonable bounds. If B is overestimated, the integral of current could return negative numbers, decreasing the accuracy greatly. For this reason, the integrating of the current will take place in time-steps, t_{step}.

Given an estimate of V_{out}, the estimated time for V_{out} to change Δ, say one volt, can be determined by

$$t_{step} = \frac{\Delta}{B}. \tag{4.14}$$

This estimate of t_{step} is then used to find the voltage change by current integration. If t_{step} was too large, the step size is decreased until the calculated voltage change is close to Δ. Given the calculated change in V_{out}, the estimate $V_{out,est}$ can be updated.

In order to control the error it is best to initially underestimate **B**. It is also necessary to control the step size. In many subcircuits, the output voltage changes little while the input is switching, and thus, the error is little. However, circuits which have large output voltage changes during the switching of an input have a greater potential for error. Thus, these circuits exercise the time-step control in order to minimize the error due to the estimate of V_{out}.

A transistor in the saturated region and the input switching according to (4.11) has the current

$$I_D = Y(V_{GSi} + S\,t - V_t)^2, \tag{4.15}$$

which reduces to

$$I_D = Y[(V_{GSi} - V_t)^2 + 2S(V_{GSi} - V_t)t + S^2t^2]. \tag{4.16}$$

Dividing through by the capacitance **C**, this is of the form

$$\frac{I_C}{C} = \alpha_1 + \alpha_2 t + \alpha_3 t^2, \tag{4.17}$$

which is integrated in (4.5) to yield the equation

$$V_{outf} - V_{outi} = \alpha_1 t_{step} + \frac{\alpha_2}{2} t_{step}^2 + \frac{\alpha_3}{3} t_{step}^3. \tag{4.18}$$

If both transistor are switching, the current equations for both devices are reduced to the form of (4.17). The current of (4.6) is then integrated to give an equation of the form of (4.18). If one transistor is switching and the other has its input fixed, as in an NMOS inverter, the device with the fixed input is also reduced to the form of (4.17). Note that (4.16) applies to the saturated region of a device with a fixed input voltage if $S = 0$.

The division by C in (4.17) can be implemented by using the term Y/C in the equations. The reason for this will become clear later when time-constants are found for the connections to V_{DD} and GND rather than keeping the Y and C terms separate. The time-constant, τ, will be applied in the equations as

$$\frac{Y}{C} = \frac{1}{\tau} \tag{4.19}$$

If a transistor is in the linear region and is switching according to (4.11), the current is

$$I_D = Y[2((V_{GSi} - V_t) + St)V_{DS} - V_{DS}^2] \tag{4.20}$$

Tho oubstitution for V_{DS} depends on which transistor the current represents. The n-type driver has $V_{DS} = V_{out}$. Substituting (4.10) for V_{out} yields

$$I_N = Y\Big[V_{outi}(2(V_{GSi}-V_{tn}) - V_{outi}) \tag{4.21}$$

$$+ 2(B(V_{GSi}-V_{tn}) + (S-B)V_{outi})t + 2B(S-B)t^2\Big].$$

For a p-type driver

$$I_P = Y\Big[(V_{outi}-V_{DD})(2(V_{GSi}-V_{tp}) - (V_{outi}-V_{DD})) \tag{4.22}$$

$$+ 2(B(V_{GSi}-V_{tp}) - (S+B)(V_{outi}-V_{DD}))t + 2B(S+B)t^2\Big].$$

After being divided by C as in (4.6), both (4.21) and (4.22) are in the form of (4.17). If both transistors supply current, the difference between the constants is used for the integration to obtain (4.18).

4.4.2 Equations for Fixed Inputs

When both transistors have inputs fixed, the equations will have a different form. For a device which is saturated,

$$I_D = Y(V_{GS} - V_t)^2, \tag{4.23}$$

which is easily integrated in (4.5) as

$$V_{outf} - V_{outi} = \frac{Y}{C}(V_{GS} - V_t)^2 t_{step} \tag{4.24}$$

or

$$t_{req} = \frac{V_{outd} - V_{outi}}{\frac{Y}{C}(V_{GS} - V_t)^2}. \tag{4.25}$$

The usage of these equations is to first find t_{req}, given the desired output voltage, V_{outd}. If the transistor is not in this state long enough, which means $t_{req} > t_{width}$, then solve for V_{outf} given the available $t_{step} = t_{width}$.

If a transistor is in the linear region, then

$$I_D = Y(2(V_{GS}-V_t)V_{DS} - V_{DS}^2). \tag{4.26}$$

Again the substitution for V_{DS} depends on which transistor the current represents. For the **n**-type driver

$$I_N = Y(2(V_{GSn} - V_{tn})V_{out} - V_{out}^2). \tag{4.27}$$

For the p-type driver

$$I_P = Y(2(V_{GSp} - V_{tp})(V_{out} - V_{DD}) - (V_{DD} - V_{out})^2). \tag{4.28}$$

This current of the form

$$\frac{I_C}{C} = \beta_1 + \beta_2 V_{out} + \beta_3 V_{out}^2 \tag{4.29}$$

is integrated in (4.5) or (4.6). If the current from the p-type device is 0, then $\beta_1 = 0$, which results in a simpler equation to integrate. If $\beta_1 = 0$, (4.29) is integrated in (4.5) to give

$$t_{req} = \frac{1}{\beta_2} \ln \left[\frac{V_{outd}}{V_{outi}} \frac{(\beta_3 V_{outi} + \beta_2)}{(\beta_3 V_{outd} + \beta_2)} \right], \tag{4.30}$$

or, if $t_{width} < t_{req}$, which means the subcircuit changes before the desired voltage is reached,

$$V_{outf} = \frac{\beta_2 \gamma}{\beta_3 (1 - \gamma)} \tag{4.31}$$

where

$$\gamma = \frac{\beta_3 V_{outi}}{\beta_3 V_{outi} + \beta_2} e^{\beta_2 t_{width}}. \tag{4.32}$$

If the p-type transistor is ON, which means $\beta_1 \neq 0$, then

$$t_{req} = \frac{1}{\beta_3} \frac{1}{\rho_2 - \rho_1} \ln \left[\frac{(V_{outd} + \rho_1)}{(V_{outi} + \rho_1)} \frac{(V_{outi} + \rho_2)}{(V_{outd} + \rho_2)} \right], \quad (4.33)$$

where ρ_1 and ρ_2 are the roots of the quadratic equation

$$z^2 + \frac{\beta_2}{\beta_3} z + \frac{\beta_1}{\beta_3} = 0 . \quad (4.34)$$

Again, if $t_{width} < t_{req}$,

$$V_{outf} = \frac{\rho_1 - \rho_2 \hat{\gamma}}{\hat{\gamma} - 1}, \quad (4.35)$$

where

$$\hat{\gamma} = \frac{V_{outi} + \rho_1}{V_{outi} + \rho_2} e^{(\rho_2 - \rho_1)\beta_3 t_{width}} . \quad (4.36)$$

Using the derived equations to find the time and voltage changes on the output node, an algorithm can be created which uses the equations to find the delay times of an output waveform. The primary use of the equations is to find the times at which the output voltage crosses V_{TL} and V_{TH}, but they must also be used to determine when the two transistors change their region of operation.

4.4.3 Using the Equations

Given the above equations, an algorithm for applying them may now be outlined. The DELAY procedure is given a dc output voltage, V_{net}, which would be reached if the configuration at the end of the phase is used to find the dc output voltage.

The current output node voltage is also specified along with the phase width, t_{width}. The time-constants τ_{VDD} and τ_{GND} are the parameters of the equivalent transistor connected to the voltage sources.

$$\frac{Y_{VDD}}{C} = \frac{1}{R_{VDD}C} = \frac{1}{\tau_{VDD}}$$

When the output voltage is falling, some parameters must be exchanged, such as the τs. The DELAY routine can handle the variables and parameters to insure the correct values are used in the equations.

The routine for switching inputs performs output voltage change calculations in time-steps. The time-step size can be controlled by the Δ parameter in Equation (4.14). The smaller Δ, the more accurate the calculation becomes. An assumption used in the switching routine is that over a time-step the output changes nearly linearly and the error induced by not changing regions of operation at the correct time in the middle of a time-step introduces little error. Thus, if the voltage change at the end of a time-step crosses a logic threshold, a linear approximation can be used to determine the time of the crossing. If the voltage change at the end of a time-step crosses a device threshold, the portion of the calculation in the wrong region is not recalculated.

```
procedure DELAY(V_net, V_out, t_width, τ_VDD, τ_GND, S_p, S_n)
begin
        if (V_net > V_TH) then
                v_tl  ← V_TL;
                v_th  ← V_TH;
                v_out ← V_out;
                v_net ← V_net;
        else
                v_tl  ← V_DD − V_TH;
                v_th  ← V_DD − V_TL;
                v_out ← V_DD − V_out;
                v_net ← V_DD − V_net;
                switch n and p parameters;
        end if
        if (S_p or S_n ∈ {SWON, SWOFF}) then
                SWITCHING(v_net, v_out, t_width, v_tl, v_th,
                                τ_VDD, τ_GND);
        else
                FIXED(v_net, v_out, t_width, v_tl, v_th, τ_VDD, τ_GND);
        end if
        if (v_net > V_TH) then
                V_out ← v_out;
        else
                V_out ← V_DD − v_out;
        end if
end
```

```
procedure SWITCHING(v_net, V_out, t_width, v_tl, v_th, τ_VDD, τ_GND)
begin
        if (there is no estimate of V_out) then
                estimate B as 0.05 volt/ns;
        end if
        while (V_out < v_net and t_width > 0) do
                if (p transistor is saturated) then
                        find α_1p, α_2p, and α_3p using (4.16)
```

```
                and (4.17);
        else if (p transistor is in the linear region) then
                find α1p, α2p, and α3p using (4.22)
                and (4.17);
        end if
        if (n transistor is saturated) then
                find α1n, α2n, and α3n using (4.16)
                and (4.17);
        else if (n transistor is in the linear region) then
                find α1n, α2n, and α3n using (4.22)
                and (4.17);
        end if
        α1 ← α1p − α1n;
        α2 ← α2p − α2n;
        α3 ← α3p − α3n;
        tstep ← min(twidth, Δ/B);
        Vouti ← Vout;
LOOP:
        find Voutf using (4.18);
        if (Voutf − Vouti > 1.5Δ) then
                decrease tstep;
                go to LOOP;
        end if
        if (Voutf > vth) then
                find threshold crossing time;
                create a transition;
                Voutf ← vnet;
        else if (Voutf > vtl and Vouti < vtl) then
                find threshold crossing time;
                create a transition;
        end if
        time   ← time + tstep;
        B      ← (Voutf − Vouti)/tstep;
        twidth ← twidth − tstep;
        Vout   ← Voutf;
    end while
end
```

```
procedure FIXED(v_net, V_out, t_width, v_tl, v_th, τ_VDD, τ_GND)
begin
        while (V_out < v_net and t_width > 0) do
                if (V_out > v_tl) then
                        V_outd ← v_th;
                else
                        V_outd ← v_tl;
                end if
                if (p transistor is saturated) then
                        find β_1p using (4.23);
                        V_outd ← min(V_outd, V_DD+V_GSp−V_tp);
                else if (p transistor is in the linear region) then
                        find β_1p, β_2p, and β_3p using (4.27)
                        and (4.29)
                end if
                if (n transistor is saturated) then
                        find β_1n using (4.23)
                else if (n transistor is in the linear region) then
                        find β_1n, β_2n, and β_3n using (4.28)
                        and (4.29)
                        V_outd ← min(V_outd, V_GSn − V_tn);
                end if
                β_1 ← β_1p − β_1n;
                β_2 ← β_2p − β_2n;
                β_3 ← β_3p − β_3n;
                find t_req using (4.25), (4.30) or (4.33);
                if (t_req > t_width) then
                        find V_outf using (4.24), (4.31) or (4.35);
                        V_out  ← V_outf;
                        time ← time + t_width;
                        t_width ← 0;
                else
                        if (V_outd = v_tl) then
                                create a transition;
                        else if (V_outd = v_th) then
```

```
                        create a transition;
                        V_outd ← v_net;
                    end if
                    V_out  ← V_outd;
                    time   ← time + t_req;
                    t_width ← t_width − t_req;
                end if
            end while
end
```

There are actually more parameters used than shown in these routines. They would be passed along with the τ's. If a transistor's input is switching, the slope (S) and the starting time of the switching are required. The threshold voltage (V_t) of the transistor and model card information, if the equations are expanded to include other parameters, are also required.

Although the above equations were derived for a CMOS inverter, the equations also apply to an NMOS inverter. The NMOS load transistor has $V_{GS} = 0$. This is the only consideration which is necessary to substitute the load transistor for the PMOS transistor.

This algorithm and the equations should provide a method for calculating delay values in a switch-level timing simulator. The following sections will go into detail about other factors which must be included in a simulator. These include the calculation of the dc output voltage, V_{net}, the reduction of a subcircuit block to the inverter primitive, and the inclusion of parasitics in the calculations.

4.5 DETERMINING THE DC OUTPUT VOLTAGE

Switch-level timing simulators first determine the dc output voltage, V_{net}, before applying any delay calculations. Thus, it is necessary that this calculation be accurate. This is particularly important for delay methods using tables or equations which give delay values. If a wrong logic state is assumed, transitions might be created which will not occur in the circuit. The delay method described in Section 4.4 is immune to this type of error because the current charging or discharging the output node is integrated. If device sizes are inappropriate, the current will not yield a sufficient change in output node voltage. Blocks with NMOS pass transistors will have a voltage drop when charged to logic 1. If this is not considered for the transistors with this reduced gate voltage, the current could be overestimated by a factor of 2 or more.

To find V_{net} all switching transistors are assumed to switch completely. A switching ON transistor is treated as an ON transistor and a switching OFF transistor is treated as OFF. Transistors with inputs in the *I* state are treated as ON transistors, although the gate voltage V_G will be reduced. The goal is to find the dc voltage after the devices have finished switching and the output has settled. There are *five* types of transistor configurations between an output node and the voltage sources V_{DD} and GND.

(1) The first configuration is when the output node is connected to only one voltage source and there are no NMOS pass transistors. In this case V_{net} takes the voltage of the voltage source. NMOS pass transistors are the special case due to the V_t voltage drop across the transistor.

(2) In the second case, the output node is connected to both V_{DD} and GND, and there are no NMOS pass transistors. Equations to determine the output state and voltage must be applied in this case. To illustrate the method, the calculation will be shown for the CMOS inverter-type gate, as shown in Figure 4.1, with both driver transistors ON. This example also applies to the NMOS inverter, where rather than a p-type transistor with V_{GSp} and V_{tp}, there is a NMOS depletion-type load transistor with $V_{GSl} = 0$ and V_{tl}. Rather than assuming the V_{GS} values for both transistors is V_{DD}, the terms V_{GSn} and V_{GSp} will be used. Thus, if either gate voltage does not swing fully, the reduced voltage is included in the calculation.

The first objective is to determine the correct dc logic state for the configuration. For each possible output voltage, there are associated currents through each transistor. The dc output voltage is that voltage where the two currents are equal. Such a voltage exists since both transistor currents are increasing functions of $|V_{DS}|$. The dc logic state is then found by comparing the transistor currents at each logic threshold V_{TL} and V_{TH}. The dc logic state is found by assuming the output voltage is V_{TH} and examining the transistor currents. If the current from V_{DD} is greater than the current from GND, then the output state is logic 1. If the current to GND is greater, then the output state is either logic 0 or I. The currents are then recalculated assuming the output voltage is V_{TL}. If the GND current is greater than the V_{DD} current, then the output state is logic 0. Otherwise, the logic output state is the intermediate state I. In the I state the voltage is taken to be $(V_{TH} + V_{TL})/2$.

Consider the CMOS gate with

$$Y_p = \frac{K_{pp}W_p}{2L_p},\ V_{tp},\ Y_n = \frac{K_{pn}W_n}{2L_n},\ \text{and } V_{tn}.$$

When the output voltage is V_{TH}, the p-type transistor is in the linear region of operation and the n-type transistor operation region depends on the relationship between $V_{GSn} - V_{tn}$ and V_{TH}. The transistor currents are found to be

$$I_n = Y_n(V_{GSn} - V_{tn})^2 \tag{4.37a}$$

or

$$I_n = Y_n[2(V_{GSn} - V_{tn})V_{TH} - V_{TH}^2] \tag{4.37b}$$

and

$$I_p = Y_p[2(V_{GSp} - V_{tp})(V_{TH} - V_{DD}) - (V_{TH} - V_{DD})^2]. \tag{4.38}$$

If I_p is greater than I_n then the output voltage is greater than V_{TH}, or logic 1. Otherwise, assume the output node voltage is V_{TL}. The n-type transistor is in the linear region of operation and the p-type transistor operation region depends on the relationship between $V_{GSp} - V_{tp}$ and V_{TL}. The currents are

$$I_n = Y_n[2(V_{GSn} - V_{tn})V_{TL} - V_{TL}^2] \tag{4.39}$$

and

$$I_p = Y_p(V_{GSp} - V_{tp})^2 \tag{4.40a}$$

or

$$I_p = Y_p[2(V_{GSp} - V_{tp})(V_{TL} - V_{DD}) - (V_{TL} - V_{DD})^2]. \quad (4.40b)$$

If I_n is greater than I_p then the output voltage is less than V_{TL}, or logic 0. Otherwise, the output voltage is between V_{TL} and V_{TH}, or logic *1*.

Having found the correct output logic state, for accuracy it is necessary to determine the actual voltage the output node will reach. This is important for switch-level timing simulation because transistor currents are sensitive to the input node voltage. It will also reduce errors in that if it is known the output voltage will not drop below, say, 0.3 volt, it is not necessary to perform the calculations to charge the output node from 0 to 0.3 when the output is switching from logic 0 to logic 1. This case occurs normally in NMOS circuits.

Given the CMOS configuration discussed above, assume the output voltage is less than V_{TL} and the actual voltage is to be found. This is determined by setting the transistor currents to be equal and solving for V_{net}, the dc output voltage.

$$I_n = Y_n[2(V_{GSn} - V_{tn})V_{net} - V_{net}^2] \quad (4.41)$$

$$I_p = Y_p(V_{GSp} - V_{tp})^2 \quad (4.42)$$

yields

$$V_{net} = (V_{GSn} - V_{tn}) - \quad (4.43)$$

$$\left[(V_{GSn} - V_{tn})^2 - \frac{Y_p}{Y_n}(V_{GSp} - V_{tp})^2\right]^{0.5}$$

This analysis is also applied when the output voltage is greater than V_{TH} or in the *I* state, although in the *I* state the voltage is taken to be $(V_{TH} + V_{TL})/2$. Consideration must be made as to the actual regions of operation of the devices. Consideration of additional model parameters would increase the accuracy, but the amount of time spent making the calculations could become prohibitive. The calculation of the dc voltage for transitions which are not complete is frequently meaningless because incomplete transitions cannot be modeled accurately as inputs to subsequent blocks.

(3) The third possible configuration is when the output node is connected to only one voltage source and there are NMOS pass transistors in the path to the output node from the source. If the source is GND, the V_{net} is also 0. If the source is V_{DD}, then the dc voltage to the output node is reduced by V_{tn} volts across the pass transistors in the path. Since the voltage drop is caused by the V_{GS} term in the transistor current equations, the voltage drop will actually be V_{tn} less than the minimum gate voltage of the pass transistors in the path. To be precise, if there are k pass transistors in the path, each with gate voltage V_{Gi}, then $V_{net} = \min(V_{Gi1}, V_{Gi2}, \ldots, V_{Gik}) - V_{tn}$.

(4) The fourth possible configuration is when the output node is connected to both voltage sources and there are NMOS pass transistors in the path to the output node from V_{DD}. The output voltage, $\hat{V}_{net}$, for the node prior to the pass transistors

can be found using (4.43). If $\hat{V}_{net}$ is less than V_{TL}, then there is no voltage drop across the pass transistors. Otherwise, the maximum voltage at the output of **k** pass transistors is the minimum pass transistor gate voltage minus V_{tn}, or $V_{net} = \min(\hat{V}_{net}, MIN_k(V_{Gi1}, V_{Gi2}, ..., V_{Gik}) - V_{tn})$.

(5) The last possible configuration is the case where the output node is disconnected from both voltage sources. In this case the output voltage stays the same unless charge sharing occurs. In the case of charge sharing, the output voltage moves towards the voltage determined by the total charge on the sharing nodes divided by the total capacitance of the nodes.

4.6 MAPPING COMPLEX BLOCKS TO PRIMITIVES

All switch-level timing simulators must have some means for mapping complex subcircuit blocks to one of the primitive blocks. After this mapping, the delay method based on that primitive is applied. If the delay method maps all transistors into linear resistors, then the theory of RC networks is applied to reduce the network to a single RC time-constant [45] or to give bounds for the response of the network [46-53]. A basis for mapping nonlinear transistors is derived in this section to show that the mapping of a subcircuit block using RC methods can be applied in a way which retains the nonlinear characteristics of the transistor.

4.6.1 Transistor Reduction Basis

Consider two transistors, m_1 and m_2, with identical process parameters and device thresholds. It is assumed that both transistors have the same gate voltage, V_G. For the transistors define

$$R_1 = \frac{1}{Y_1} = \frac{2L_1}{K_p W_1}$$

and

$$R_2 = \frac{1}{Y_2} = \frac{2L_2}{K_p W_2}$$

to be the resistances of the transistors as defined in Sections 4.3 and 4.4. The current through the transistors depends on the region of operation of the device, but is of the form

$$I = \frac{1}{R} f(V_{GS}, V_{DS}),$$

ignoring the dependence of V_t on V_{SB} (body effects). We find equivalent transistors for parallel and series combinations of the two transistors, and equivalent R_p and R_s values for the parallel and series combinations, respectively.

When the two transistors are in a parallel configuration, the current through the devices is the sum of the individual currents. Since gate, source, and drain voltages are equal, the net current is

$$I = [\frac{1}{R_1} + \frac{1}{R_2}] f(V_{GS}, V_{DS})$$

or

$$I = \frac{1}{R_p} f(V_{GS}, V_{DS}),$$

where

$$R_p = \frac{R_1 R_2}{R_1 + R_2}. \tag{4.44}$$

Thus, parallel transistors with identical gate voltages are combined into one equivalent transistor with resistance R_p.

When the two transistors are in series, as shown in Figure 4.3, the transistors are mapped by equating the transistor currents. It is assumed that the common node, V_n, has negligible capacitance. The equivalent current depends on the regions of operations of the two transistors. First the currents are set equal to find the internal node voltage. This voltage is then used to find the value of the current. The following illustrates the analysis for m_1 in the linear region and m_2 saturated. Assume m_1 and m_2 are p-type transistors with V_t, both have gate voltages of V_G, and m_1 is also connected to V_{DD}.

If m_1 is in the linear region and m_2 is saturated, the device currents are

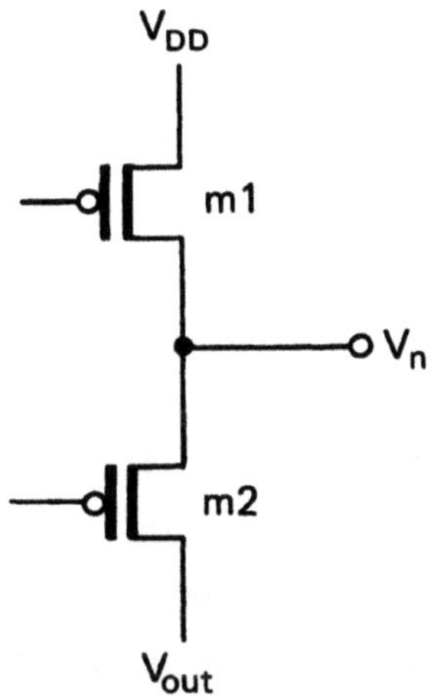

Figure 4.3: Transistor configuration for combination

$$I_1 = \frac{1}{R_1}[2(V_G - V_{DD} - V_t)(V_n - V_{DD}) - (V_n - V_{DD})^2] \quad (4.45)$$

and

$$I_2 = \frac{1}{R_2}(V_G - V_n - V_t)^2. \quad (4.46)$$

Equating the currents and solving for V_n yield

$$V_n = (V_G - V_t) + (V_G - V_{DD} - V_t)\left[\frac{R_2}{R_1 + R_2}\right]^{0.5}. \quad (4.47)$$

Substituting the value of V_n back into (4.46) yields

$$I_2 = \frac{1}{R_1 + R_2}(V_G - V_{DD} - V_t)^2, \tag{4.48}$$

which is the amount of current which flows through an equivalent transistor connected between V_{DD} and V_{out} with resistance R_s, where

$$R_s = R_1 + R_2. \tag{4.49}$$

This analysis can be repeated for both transistors in the linear region. The cases of both transistors saturated or m_1 saturated and m_2 in the linear region cannot occur in actual circuits if the gate voltages of the devices are equal. In these cases the calculated V_n contradicts the relationships required for the assumed regions of operation of the devices. The analysis assumes the internal node capacitance is negligible, which might not be true for small devices. This provides a basis for approximation methods in analyzing subcircuit blocks. There are two methods which are used to find an equivalent time-constant through a network of transistors with a grounded capacitance at every node.

(1) All internal capacitances may be either ignored or lumped to the output node. A graph is constructed where an edge with its associated conductance is created for each transistor in the block. An equivalent conductance from the voltage source to the output node is then found by systematically eliminating all internal nodes from the graph. The output node capacitance is then divided by this conductance to obtain a time-constant for the path. This time-constant is then used in the voltage equations as shown in

(4.19). This technique will be discussed in more detail later.

(2) RC network theory can also be used to process the network through matrix or tree techniques [45-53]. The matrix technique may be more time-consuming than the first approach because of the addition of the internal capacitances. The analysis can be simplified if the network is a tree structure. The Elmore time-constant [45] is easily found for an RC-tree. The advantage of these methods is that they find an equivalent RC time-constant, τ, for the network which models the internal node capacitances more accurately than the first method.

The first approach above has the advantage of simplicity, but the second approach has an advantage in that a time-constant is found, rather than an equivalent resistance. This is particularly important for circuits which contain pass transistors. The capacitance of nodes connected to pass transistors should not be ignored or lumped to the output. A combination of the above methods which first eliminates internal nodes with negligible capacitance and then finds a time-constant for the reduced network can result in a method which is both fast and accurate.

Equations (4.6) and (4.19) show the fact that capacitance is part of the delay equations in the form of an RC time-constant. Thus, finding a time-constant for a network can be directly applied in the equations. The reduction will contain two steps: 1) reduce the subcircuit as much as possible by eliminating internal nodes with small capacitance, 2) parse the reduced subcircuit to obtain an Elmore time-constant for the output node. This time-constant may then be applied in the equations for the terms of

$$\frac{Y}{C} = \frac{1}{RC} = \frac{1}{\tau}. \tag{4.50}$$

4.6.2 Subcircuit Reduction Algorithm

The objective of the reduction algorithm is to first reduce the subcircuit graph according to the first step above, then parse the reduced graph according to step two to find time-constants with which to apply the voltage equations. Most combinational gates consist of only series-parallel transistor connections. Thus, most subcircuits will be reduced to the inverter case in the first step itself if internal node capacitances are ignored. The second step is applied to those circuits with some large internal capacitance, such as pass transistor networks.

Consider a subcircuit block Ω_q in the network and a phase ϕ of simulation. Let **OUT**, **INTERNAL**, **H** etc. be as defined in Section 4.3. Let $C_{min} = \min_{n \in OUT} CAP(n)$ denote the minimum capacitance of the output nodes of the block. Let **ELIM** denote the set of vertices in **H** defined to be those vertices corresponding to nodes in **INTERNAL** with capacitances less than $\delta\, C_{min}$ where δ is some small fraction which, in our case, is chosen to be 0.15. Nodes with capacitances less than $\delta\, C_{min}$ are assumed to have negligible effect on the charging of the output nodes because 1) the capacitance to be charged is dominated by the output node, and 2) the resistance of the path from a voltage source to the output node is greater than the resistance of a path to an internal node. These two factors indicate the time-constant for the output node is dominated by the output node itself if internal nodes have capacitances less than $\delta\, C_{min}$.

A *simple graph* is a graph with no self-loops and no parallel edges. The *simplification of a graph* is a new graph obtained by collapsing all parallel edges into a single edge. Let $e_1, e_2, ..., e_k$ be a set of parallel edges in a graph with conductances $Y_1, Y_2, ..., Y_k$ and states $S_1, S_2, ..., S_k$. The new edge will have an edge conductance Y_{new} and state S_{new} defined as follows:

(1) The ordering of the states is assumed to be

$$\mathrm{ON} > \mathrm{SWON} > \mathrm{INT} > \mathrm{SWOFF} > \mathrm{OFF} \tag{4.51}$$

meaning a transistor which is **ON** has greater current than a transistor which is **SWON**. A transistor which is **SWOFF** will have less current than a transistor in the **INT** state.

(2) The state of the new edge is defined as

$$S_{new} = \mathrm{MAX}\,\{\, S_1, S_2, ..., S_k \,\},$$

where the **MAX** operation is defined according to the above ordering given in Equation (4.51). This means the current through the equivalent transistor will behave like the current through the strongest of the individual transistors.

(3) Define an indicator set of indices $\Gamma = \{\, i : S_i = S_{new} \,\}$. Then the conductance of the new edge is

$$Y_{new} = \sum_{i \in \Gamma} Y_i \tag{4.52}$$

We now eliminate all the internal vertices in the graph since the capacitance at the corresponding nodes is negligible. Let $v \in$ ELIM denote a vertex corresponding to an internal node with a relatively small capacitance. Let $w_1, w_2, ..., w_k$ denote the vertices adjacent to v in the graph and let Y_i denote the conductance of the edge $\langle v,w_i \rangle$ whose state is S_i. Also, let $Y_{tot} = \sum_{i=1}^{k} Y_i$ denote the sum of all conductances connected to v. We define the *elimination of a vertex* v *from a simple* graph as a procedure involving the following two steps :

(1) For every pair of vertices w_i and w_j adjacent to v in the graph, add an edge e_{ij} between w_i and w_j. The conductance of this new edge added is defined to be

$$Y_{ij} = \frac{Y_i\, Y_j}{Y_{tot}}. \tag{4.53}$$

The state of this new edge added is $S_{ij} = \mathrm{MIN}\{ S_i, S_j \}$, where the MIN operation is performed relative to the ordering of the states as given by (4.51).

(2) Delete the vertex v (and all edges incident on it) from the new graph obtained in step (1).

It must be noted that eliminating a vertex from a simple graph could create parallel edges in the new graph. Hence, the process of elimination should be followed by simplification of the graph. The procedure REDUCE used to eliminate the vertices in ELIM is outlined below.

```
procedure REDUCEGRAPH (H)
begin
        H←SIMPLIFY(H);
        for each vertex v ∈ ELIM do
                H ← ELIMINATE(v, H);
                H ← SIMPLIFY(H);
        end for
end

procedure ELIMINATE (v, H)
begin
        Ĥ    ← H
        Y_tot ← 0.0 ;
        for each vertex w ∈ Adj_H(v) do
                e     ← <v, w> ;
                Y     ← 1.0/RES(e) ;
                Y_tot ← Y_tot + Y ;
        end for
        for each pair w_i, w_j ∈ Adj_H(v) do
                e_i ← <v, w_i> ;
                e_j ← <v, w_j> ;
                Y_i ← 1.0/RES(e_i) ;
                Y_j ← 1.0/RES(e_j) ;
                add a new edge e_ij in Ĥ joining w_i and w_j ;
                RES(e_ij) ← Y_tot / (Y_i·Y_j) ;
                STATE(e_ij) ← MIN{STATE(e_i),
                                  STATE(e_j)};
        end for
        return Ĥ−v;
end
```

In the above procedure, the choice of v as an internal vertex picked for elimination from **H** is important from the complexity point of view. If the degree of v in **H** is **k** then the total number of edges added as a result of eliminating v from **H** is $k(k-1)/2 - k$ which is equal to $k(k-3)/2$. If $k = 2$ then the new graph has one edge less than the number in **H** while the number of edges is unchanged if $k = 3$. Hence, a vertex of lowest degree in **H** is picked as the best candidate for elimination.

We now review the notion of the Elmore time-constant [46] in an RC-tree. A graph T is a *tree* if it is connected and has no cycles. In each tree, we can focus our attention on a special vertex called the *root* of the tree. If a vertex a is a root of a tree T, then T is said to be rooted at a, denoted by T_a. In any tree, there is a unique path from the root to any other vertex in the tree (in fact, there is a unique path between any two vertices in a tree). We say that a network composed of resistances and capacitances forms an *RC-tree* if the subnetwork of resistances, when viewed as a weighted graph, forms a tree and there is a capacitance from each node of the network to ground. Note that all capacitors in such a network are grounded, i.e., there are no floating capacitors. Consider an RC-tree rooted at node n_0 and let $n_1, n_2, ..., n_p$ be the rest of the nodes. Let C_i denote the capacitance from node n_i to ground, for each $i = 1, 2, ..., p$. Let P_i denote the unique path from the root n_0 to the node n_i and let $P_{ij}=P_i \cap P_j$ denote the portion of the path between the root and n_i that is common to that between the root and n_j. Let R_{ij} denote the sum of all the resistances in P_{ij}. If $P_{ij}=\emptyset$, then $R_{ij}=0$. We can now associate a time-constant τ_i, known as *Elmore time-constant* for each node n_i in the RC-tree, defined as:

$$\tau_{Elm,i} = \sum_{j=1}^{p} R_{ij} C_j. \tag{4.54}$$

An example of an RC-tree is shown in Figure 4.4. For the node n_1 in the network in Figure 4.4, the Elmore time-constant is

$$\tau_{Elm,1} = R_1(C_1+C_2+C_3+C_4+C_5+C_6+C_7)$$

while for node n_7 it is

$$\tau_{Elm,7} = R_1(C_1+C_2+C_3+C_4+C_5+C_6+C_7) +$$

$$R_3(C_3+C_4+C_6+C_7) + R_4(C_4+C_7) + R_7C_7.$$

Let us now go back to the simulation of a block. In most cases the procedure REDUCE outlined above will reduce the original graph **H** to a tree structure (or a disjoint union of trees). We then construct an RC-tree from the reduced graph by replacing each edge by a resistance equal to the resistance of the corresponding transistor and a capacitance = **CAP(n)** from each node **n** to ground. For each output node in **OUT** in this tree we obtain two time-constants τ_{VDD} and τ_{GND} as follows. First, we label the vertex corresponding to V_{DD} as a root and delete the **GND** node from the network. We then compute τ_{VDD} to be the Elmore time-constant for the output node under consideration. This process is repeated by labeling the **GND** node as the root and deleting the V_{DD} node from the network. The resulting Elmore time-constant is τ_{GND}. These time-constants are then applied in the delay equations as shown in (4.19).

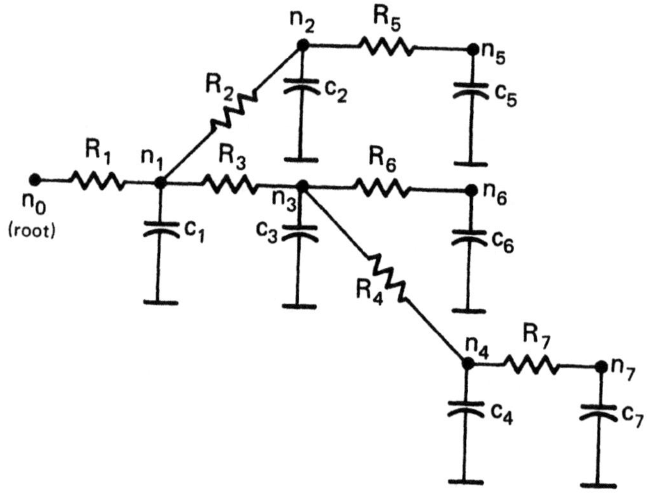

Figure 4.4: An RC-tree rooted at n_0

If the reduced graph is not a tree, some approximation must be used to process the graph. A simple approach is to find a tree which spans the graph, and use that tree to find the time-constant. By a spanning tree of a graph, we mean, a subgraph which is a tree and includes all the nodes of the original graph. The tree is found by finding the least resistance path to each node. This approximation uses the assumption that a node will be charged or discharged by the least resistance path from that node to the voltage source.

The following PARSEGRAPH procedure finds a spanning tree using the recursive procedure SEARCH1. The Elmore time-constant is then found by parsing the tree and accumulating the terms for each node. The parsing of the tree is performed by the recursive procedure SEARCH2.

```
procedure PARSEGRAPH(H, output, source, τ)
begin
        for each vertex n ∈ H do
                onpath(n)  ← FALSE;
                onstack(n) ← FALSE;
                parent(n)  ← NIL;
                rtosrc(n)  ← ∞;
                state(n)   ← OFF;
        end for
        onpath(source)  ← TRUE;
        onstack(source) ← TRUE;
        rtosrc(source)  ← 0;
        state(source)   ← ON;
        SEARCH1(H, source);
        n ← output;
        while n≠source do
                onpath(n) ← TRUE;
                n ← parent(n);
        end while
        τ ← 0;
        SEARCH2(H, source, τ, 0);
end
```

```
procedure SEARCH1(H, v)
begin
        if (state(v)=OFF)
                return;
        for each vertex n ∈ Adj_H(v) do
                if (onstack(n))
                        continue;
                e      ← <v, n>;
                r      ← rtosrc(v) + RES(e);
                state ← MIN{state(v), STATE(e)};
                if (state<state(n))
                        continue;
                else if (state = state(n))
                        if (r > rtosrc(n))
                                continue;
                state(n)    ← state;
                rtosrc(n)   ← r;
                parent(n)   ← v;
                onstack(n) ← TRUE;
                SEARCH1(H, n);
                onstack(n) ← FALSE;
        end for
end


procedure SEARCH2(H, v, tau, rparent)
begin
        r_ij ← rparent;
        if (onpath(v))
                r_ij ← rtosrc(v);
        τ ← τ + r_ij CAP(v);
        for each vertex n ∈ Adj_H(v) do
                if (parent(n) = v)
                        SEARCH2(H, n, τ, r_ij);
end
```

4.7 PARASITICS

With the decreasing transistor dimensions the delay due to parasitics has become an important aspect of switch-level timing simulation. It is, therefore, necessary for the switch-level timing simulator to be capable of including the parasitics in the delay method. A subcircuit block is viewed as a connection of transistors followed by RC networks on the output nodes of the transistor network, as shown in Figure 4.3. Delay methods which model the transistor network as an RC network have no difficulty in including the parasitics. Unfortunately, transistors are not linear resistors. A simple method for including parasitics in the delay method described in Section 4.4 is to model the parasitic resistances as nonlinear resistances and combine the RC network with the transistor network using the methods described in Section 4.6. This type of mapping may result in greater overall accuracy, since the current charging the output node is controlled primarily by the nonlinear transistors.

The subcircuit block reduction method described in Section 4.6 essentially calculates an equivalent Elmore time-constant for the transistor network, and then applies the nonlinear equations using the resulting time-constant. This leads to a simple method for combining the parasitics with the derived time-constant. The RC parasitic tree is parsed to find Elmore time-constants for each output node. The parasitic time-constant is then combined with the transistor subcircuit time-constant to give a net time-constant for use in the equations. The only requirement is that a mapping is needed to scale the parasitic time-constant so that when the nonlinear equations are applied, the current through a parasitic resistor is approximated correctly. This scale factor is derived by equating the resistor current with that of the equivalent transistor model.

Given a linear resistor R, the equivalent nonlinear transistor resistance $\hat{R}$ is to be found. When a transistor is in the saturation region, the current is

$$I_D = \frac{1}{\hat{R}}(V_{GS} - V_t)^2, \tag{4.55}$$

while the resistor current is $I_R = \frac{V}{R}$. Setting the currents equal and $V = V_{DS}$ yields

$$\hat{R} = R\frac{(V_{GS} - V_t)^2}{V_{DS}}. \tag{4.56}$$

If the transistor is ON and $V_{DS} = V_{DD}$, then

$$\hat{R} = R\frac{(V_{DD} - V_t)^2}{V_{DD}}. \tag{4.57}$$

If $V_{DD} = 5.0$ and $V_t = 1.0$, then $\hat{R} = 3.2$ R. When a transistor is in the linear region, the current is

$$I_D = \frac{1}{\hat{R}}[2(V_{GS} - V_t)V_{DS} - V_{DS}^2], \tag{4.58}$$

which yields

$$\hat{R} = R\frac{[2(V_{DD} - V_t)V_{DS} - V_{DS}^2]}{V_{DS}}. \tag{4.59}$$

If $V_{DS} = 4.0, V_{DD} = 5.0$, and $V_t = 1.0$, then $\hat{R} = 4.0\,R$. If $V_{DS} = 3.0$, then $\hat{R} = 5.33\,R$. If $V_{DS} = 2.0$, then $\hat{R} = 6.0\,R$.

Based on this type of analysis an appropriate scaling factor is selected to convert the parasitic time-constant to that of the transistor time-constant. The scaling factor should try to model all forms of the calculations of $\hat{R}$. A good scaling factor will be relatively accurate over a wide range of operating points. The primary application is for larger values of V_{DS}. A possibility is to average (4.57) and (4.59) using $V_{DS} = V_{DD}/2$. This would result in

$$K_R = \frac{\hat{R}}{R} = \frac{(V_{DD} - V_t)^2}{2V_{DD}} + (V_{DD} - V_t) - \frac{V_{DD}}{4}. \tag{4.60}$$

Consider an RC network being driven by a subcircuit block of transistors. Let $C_1, C_2, \ldots, C_m$ be the capacitances in the RC network and let τ_r denote the Elmore time-constant between the root and some output node of the RC network. Let τ_{VDD} and τ_{GND} denote the voltage source time-constants from the output of the subcircuit block, and let R_{VDD} be the equivalent "resistance" from V_{DD} to the output node of the block and R_{GND} is the equivalent "resistance" from GND to the output node of the block. In fact, τ_{VDD}, τ_{GND}, R_{VDD}, and R_{GND} are set by calls to PARSEGRAPH. The new τ^*_{VDD} and τ^*_{GND} are then found as

$$\tau^{*}_{VDD} = \tau_{VDD} + R_{VDD}(C_1 + C_2 + \cdots + C_m) + \tau_r K_R \quad (4.61)$$

$$\tau^{*}_{GND} = \tau_{GND} + R_{GND}(C_1 + C_2 + \cdots + C_m) + \tau_r K_R. \quad (4.62)$$

K_R converts the linear parasitic RC time-constant to the scale of the current equations. These new time-constants are then used in the delay calculations.

4.8 SAMPLE SUBCIRCUIT PROCESSING

To illustrate the techniques described in this chapter, some examples will be described. The first example will demonstrate the application of the delay equations on a simple CMOS inverter. The next example of a NAND gate will illustrate the concepts used in reducing a block. The parsing of a block with more than one output node will be demonstrated using an NMOS inverter driving a pass transistor.

4.8.1 Simple CMOS Inverter

The processing of a simple CMOS inverter is described in order to explain the phases used in the simulation process and the delay equations. Many circuit parameters, the τ's in particular, will not be given in order to concentrate on the type of processing. The following values are used in the processing: $V_{DD} = 5.0$ volts, $V_{TL} = 1.0$ volt, $V_{TH} = 4.0$ volts, $V_{tn} = 1.0$ volt, $V_{tp} = -1.0$ volt, and $\Delta = 1.0$ volt. The CMOS inverter shown in Figure 4.5 has the following input transition sequence:

$$(0.2 \text{ ns}, 0 \rightarrow I)(0.8 \text{ ns}, I \rightarrow 1, 5.0)$$

$$(11.2 \text{ ns}, 1 \rightarrow I)(11.8 \text{ ns}, I \rightarrow 0, 0.0).$$

This input sequence is then used to find the phases for processing. The phase ending time for the switching input has been modified. Thus, the calculations for the switching input continues until the input voltage reaches its stable value. During each phase, each transistor has a fixed state. For convenience, the phases are numbered here.

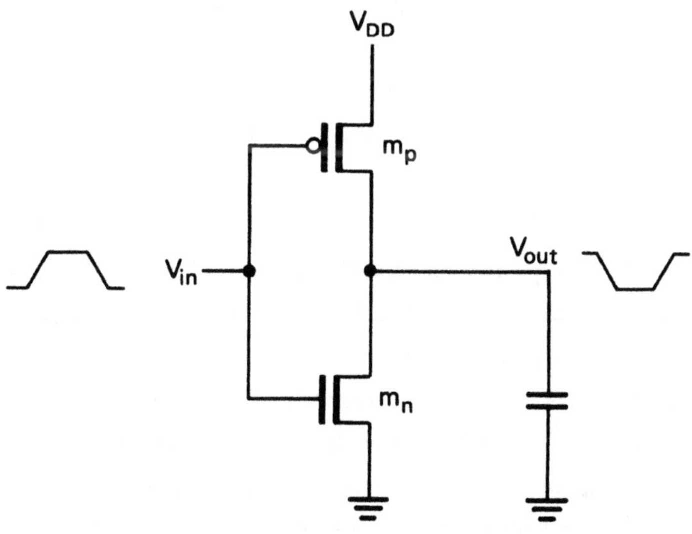

Figure 4.5: CMOS inverter for example 4.8.1

1 (0.0 ns – 0.2 ns, m_p is ON, m_n is OFF)

2 (0.2 ns – 1.0 ns, m_p is SWOFF, m_n is SWON)

3 (1.0 ns – 11.2 ns, m_p is OFF, m_n is ON)

4 (11.2 ns – 12.0 ns, m_p is SWON, m_n is SWOFF)

5 (12.0 ns – ∞, m_p is ON, m_n is OFF)

The initial state of the output node is determined by finding V_{net} for the first phase. Initially, the output voltage is 5.0 volts. The connections to V_{DD} and GND are found by the PARSEGRAPH routine as each phase is processed. If there is no connection to a voltage source, τ is taken as ∞. This is equivalent to a transistor with width 0. Since the output voltage for the first phase equals the V_{net}, no processing is necessary.

In the second phase, the mapping procedure finds the SWOFF path to V_{DD} and the SWON path to GND. The calculation of V_{net} assumes the transistors have finished switching, leaving only a connection to GND. Thus, $V_{net} = 0.0$ volt and delay calculations will be performed. The objective is to find the times t_1 and t_2 when the output voltage crosses the logic thresholds V_{TH} and V_{TL}, respectively. The delay calculations are derived for a rising output voltage. In this case the output voltage is falling, so V_o is used here as the output voltage used in the equations, which is rising and $V_{out} = V_{DD} - V_o$.

For this phase the initial value of t_{width} is 0.8 ns. Following the SWITCHING routine, B is estimated to be 0.05 volt/ns.

m_p is in the linear region and m_n is initially saturated. $t_{step} = 0.28$ ns and $V_{outi} = 0.0$. Using (4.18) V_{outf} is found to be 0.0312 volt, so the following values are updated: $B \leftarrow 0.111$ volt/ns, $V_o \leftarrow 0.0312$ volt, and $t_{width} \leftarrow 0.52$ ns.

The regions of operations of the devices remain the same during the second pass through the loop. t_{step} is set to 0.52 ns because the remaining width of the phase is smaller than the estimated time for the output to change Δ volts, and $V_{outi} \leftarrow 0.0312$ volt. Using (4.18), V_{outf} is found to be 0.844 volt, and the following values are updated: $B \leftarrow 1.53$ volts/ns, $V_o \leftarrow 0.844$ volt, and $t_{width} \leftarrow 0.0$ ns. Since the end of the phase is reached, the processing moves on to the third phase.

The third phase finds only a connection to GND. Thus, the FIXED routine is entered with $V_o = 0.844$ volt, $V_{outd} = 1.0$ volt, and $t_{width} = 10.2$ ns. m_p is OFF and m_n is saturated, so t_{req} is found using (4.23b) to be 0.054 ns and the following updates are made: $V_o \leftarrow 1.0$ volt, $t_{width} \leftarrow 9.66$ ns, and a transition is created with a crossing time of $t_1 = 1.054$ ns. V_{outd} is updated to 4.0 volts and m_n is now in the linear region. t_{req} is now found using (4.31) to be 1.274 ns and the following updates are made: $V_o \leftarrow 5.0$ volts, $t_{width} \leftarrow 8.386$ ns, and a transition is created with a crossing time of $t_2 = 2.328$ ns. Since both threshold crossing times are found, $V_{out} \leftarrow V_{net}$ and the delay calculations are completed for this phase.

After the mapping has been found for the fourth phase, the switching OFF path to GND and the switching ON path to V_{DD} are found. Thus, $V_{net} = 5.0$ volts and delay calculations will be performed. The objective is to find the times t_1 and t_2 when the output voltage crosses the logic thresholds V_{TL} and V_{TH} respectively. The delay calculations are derived for a rising output voltage, which occurs in this case.

For this phase t_{width} is set to 0.8 ns. The switching routine estimates **B** to be 0.05 volt/ns. m_n is in the linear region and m_p is initially saturated. $t_{step} = 0.28$ ns and $V_{outi} = 0.0$ volt. Using (4.18), V_{outf} is found to be 0.0861 volt, so the following values are updated: $\mathbf{B} \leftarrow 0.307$ volt/ns, $V_{out} \leftarrow 0.0861$ volt, and $t_{width} \leftarrow 0.52$ ns.

The regions of operations of the devices remain the same during the second pass through the loop. t_{step} is set to 0.52 ns because the remaining width of the phase is smaller than the estimated time to change Δ volts, and $V_{outi} \leftarrow 0.0861$ volt. Using (4.18), V_{outf} is found to be 2.12 volts, which is too large a change, so t_{step} is reduced to 0.391 ns. Again using (4.18) V_{outf} is found to be 1.19 volts, which crosses a logic threshold. The threshold crossing time is found through interpolation as $t_1 = 11.806$ ns which was 0.326 ns after the last step. The following values are updated: $\mathbf{B} \leftarrow 2.83$ volts/ns, $V_{out} \leftarrow 1.0$ volt, and $t_{width} \leftarrow 0.194$ ns. Both transistors are now in the linear region. $t_{step} \leftarrow 0.194$ ns and $V_{outi} \leftarrow 1.0$ volt. Equation (4.18) finds V_{outf} to be 2.23 volts, and $t_{width} \leftarrow 0.0$ ns. Since the end of the phase is reached, the processing moves on to the fifth phase.

The fifth phase finds only a connection to V_{DD} and the FIXED routine is entered with $V_{out} = 2.23$ volts, $V_{outd} = 4.0$ volts, and $t_{width} = \infty$. m_p is in the linear region and m_n is OFF, so t_{req} is found using (4.31) to be 0.345 ns and the following updates are made: $V_o \leftarrow 5.0$ volts, $t_{width} \leftarrow \infty$, and the final transition is created with a crossing time of 12.345 ns. Since both threshold crossing times are found, $V_{net} = V_{out}$ and the delay calculations are completed for this phase.

Having completed all phases, the output transition sequence of the inverter is

$$(1.054 \text{ ns}, 1 \rightarrow I)(2.328 \text{ ns}, I \rightarrow 0, 0.0)$$

$$(11.806 \text{ ns}, 0 \rightarrow I)(12.345 \text{ ns}, I \rightarrow 1, 5.0).$$

The sequence generated by finding the threshold crossing time for the output waveform generated by SPICE2 is

$$(1.037 \text{ ns}, 1 \rightarrow I)(2.379 \text{ ns}, I \rightarrow 0, 0.0)$$

$$(11.798 \text{ ns}, 0 \rightarrow I)(12.367 \text{ ns}, I \rightarrow 1, 5.0).$$

which is in excellent agreement with the earlier sequence, especially, regarding the SLT estimates.

4.8.2 CMOS NAND Gate

In order to illustrate the subcircuit block reduction, a NAND gate, as shown in Figure 4.6, will be reduced to an equivalent inverter. The primary objective of this example is to demonstrate the series and parallel combination of transistors during the mapping of a block to an inverter. The internal node has capacitance $< \delta C_{out}$. The input V_1 has the transition sequence

$$(0.2\text{ ns}, 0 \rightarrow I)(0.8\text{ ns}, I \rightarrow 1, 5.0)(31.2\text{ ns}, 1 \rightarrow I)$$

$$(31.8\text{ ns}, I \rightarrow 0, 0.0)(45.2\text{ ns}, 0 \rightarrow I)(45.8\text{ ns}, I \rightarrow 1, 5.0)$$

and V_2 has the transition sequence

$$(15.2\text{ ns}, 0 \rightarrow I)(15.8\text{ ns}, I \rightarrow 1, 5.0)(30.2\text{ ns}, 1 \rightarrow I)$$

$$(30.8\text{ ns}, I \rightarrow 0, 0.0)(45.2\text{ ns}, 0 \rightarrow I)(45.8\text{ ns}, I \rightarrow 1, 5.0)$$

$$(48.2\text{ ns}, 1 \rightarrow I)(48.8\text{ ns}, I \rightarrow 0, 0.0).$$

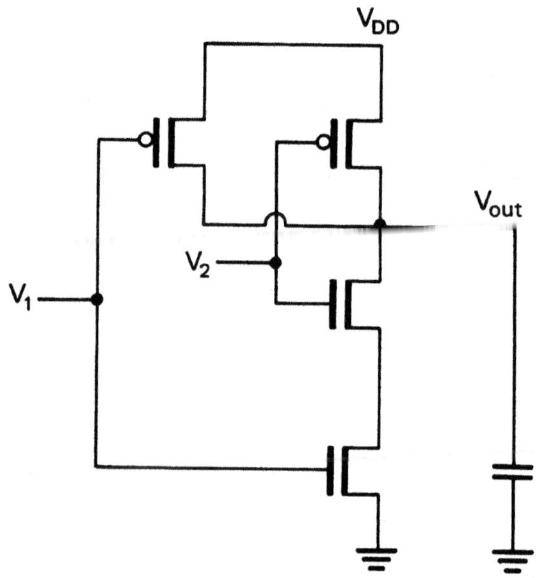

Figure 4.6: CMOS NAND gate for example 4.8.2

During the various phases of simulation, different combinations of series and parallel transistors will arise. The resistances will be combined as described in Section 4.6. The notation which will be used here is: R_1 is the resistance of the PMOS transistor with gate node V_1, R_2 is the resistance of the PMOS transistor with gate node V_2, R_p is the parallel resistance of the two PMOS transistors as defined in (4.44), R_s is the series resistance of the two NMOS transistors as defined in (4.49). The phases as constructed in Section 4.8.1 will be used again with the addition of two parameters. In addition to the state of each equivalent transistor, the resistance of the equivalent transistor will be given. The phases produced by the above input waveforms are

1 (0.0 ns – 0.2 ns, m_p is ON–R_p, m_n is OFF)

2 (0.2 ns – 1.0 ns, m_p is ON–R_2, m_n is OFF)

3 (1.0 ns – 15.2 ns, m_p is ON–R_2, m_n is OFF)

4 (15.2 ns – 16.0 ns, m_p is SWOFF–R_2, m_n is SWON–R_s)

5 (16.0 ns – 30.2 ns, m_p is OFF, m_n is ON–R_s)

6 (30.2 ns – 31.0 ns, m_p is SWON–R_2, m_n is SWOFF–R_s)

7 (31.0 ns – 31.2 ns, m_p is ON–R_2, m_n is OFF)

8 (31.2 ns – 32.0 ns, m_p is ON–R_2, m_n is OFF)

9 (32.0 ns – 45.2 ns, m_p is ON–R_p, m_n is OFF)

10 (45.2 ns – 46.0 ns, m_p is SWOFF–R_p, m_n is SWON–R_s)

11 (46.0 ns – 48.2 ns, m_p is OFF, m_n is ON–R_s)

12 (48.2 ns – 49.0 ns, m_p is SWON–R_2, m_n is SWOFF–R_s)

13 (49.0 ns – ∞, m_p is ON–R_2, m_n is OFF)

In phase 2 the state of the connections to both voltage sources remains the same, but the strength of the connection to V_{DD} has changed because one PMOS transistor is now SWOFF. Since the ON transistor has a stronger state, the equivalent transistor has the state and resistance of the single transistor. In all cases the equivalent resistance of the series NMOS transistors remains the same, and the state of the equivalent transistor is the state of the weaker transistor. Although the parameters are the same during phases 2 and 3, the phases remain separate because the block configuration is different in the two phases. During phase 10 both inputs switch at the same time; thus, the equivalent resistance will account for this case. The output sequence generated by the SIMULATE procedure is

(16.6 ns, $1 \rightarrow I$)(19.8 ns, $I \rightarrow 0$, 0.0)(31.2 ns, $0 \rightarrow I$)

(32.9 ns, $I \rightarrow 1$, 5.0)(46.6 ns, $1 \rightarrow I$)(49.9 ns, $I \rightarrow 1$, 5.0).

During phases 10 and 11 the output voltage changed approximately 2.0 volts according to the delay calculations. During phases 12 and 13 the output node was recharged, creating an incomplete transition The transition sequence as determined from the SPICE2 output waveform is

$$(16.6\text{ ns}, 1 \rightarrow I)(19.6\text{ ns}, I \rightarrow 0, 0.0)(31.3\text{ ns}, 0 \rightarrow I)$$

$$(32.7\text{ ns}, I \rightarrow 1, 5.0)(46.5\text{ ns}, 1 \rightarrow I)(50.8\text{ ns}, I \rightarrow 1, 5.0).$$

which, once again, illustrates the excellent accuracy of the SLT estimates.

4.8.3 NMOS Inverter Driving a Pass Transistor

The application of the Elmore time-constant to map a multiple output gate to that of a single inverter will be illustrated here. The circuit used in this example is shown in Figure 4.7. In this case, if the resistance of the pass transistor is R_n, the resistance of the driver transistor is $0.5R_n$ and the resistance of the load device is $2R_n$. The inverter input transition sequence is

$$(1.6\text{ ns}, 1 \rightarrow I)(2.6\text{ ns}, I \rightarrow 0, 0.0)$$

and the control input sequence is

$$(10.6\text{ ns}, 0 \rightarrow I)(11.6\text{ ns}, I \rightarrow 1, 5.0).$$

Initially the pass transistor is OFF when the inverter input switches. In this case the delay for V_{out1} is found in the same manner as in example 4.8.1.

$$\tau_{1,VDD} = R_n\,(2C_1)$$

$$\tau_{1,GND} = R_n \left(\frac{C_1}{2}\right)$$

After V_{out1} is charged, the pass transistor switches ON. The

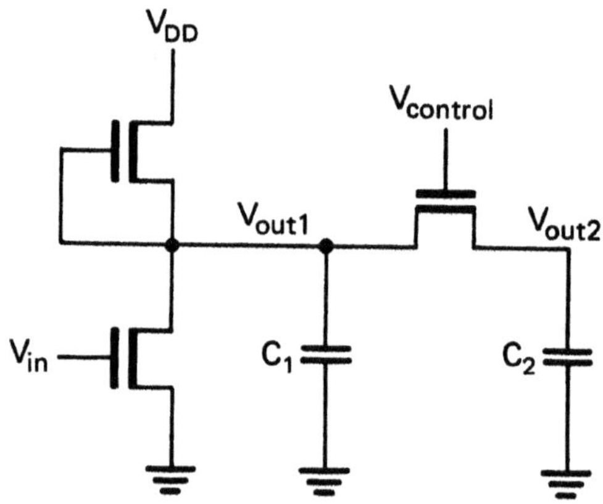

Figure 4.7: NMOS inverter and pass transistor for example 4.8.3

time-constants for the output nodes when the pass transistor switches ON is

$$\tau_{1,VDD} = R_n\, 2(C_1 + C_2)$$

$$\tau_{1,GND} = R_n \left(\frac{C_1 + C_2}{2} \right)$$

$$\tau_{2,VDD} = R_n (2C_1 + 3C_2)$$

$$\tau_{2,GND} = R_n \left(\frac{C_1}{2} + 3C_2 \right).$$

The load transistor is always ON; thus, V_{out1} will always have a connection to V_{DD}. When the driver transistor is ON, there is a connection to both V_{DD} and GND, and thus, an accurate calculation of V_{net} must be made to determine the output voltage. In this case (4.43) yields a dc voltage of 0.29 volt, which is also the dc voltage obtained in SPICE2. Since the driver transistor is already OFF when the pass transistor switches ON, there is no connection to GND for either node in this case. Another consideration for this circuit is the voltage drop across the pass transistor. Thus, V_{net} for V_{out2} is only 4.0 volts when the driver transistor is OFF and the pass transistor is ON. For this reason, V_{TH} is set to 3.5 volts when simulating this circuit, and all transition sequences in this example use this threshold value.

The output sequences produced by the SIMULATE procedure are

$$V_{out1}\text{: } (2.3 \text{ ns}, 0 \rightarrow I)(3.6 \text{ ns}, I \rightarrow 1, 5.0)$$

$$V_{out2}\text{: } (11.2 \text{ ns}, 0 \rightarrow I)(15.0 \text{ ns}, I \rightarrow 1, 4.0)$$

The Elmore time-constant assumes all nodes in the tree are initially discharged, which is not the case here. V_{out1} will experience a temporary drop in voltage when the pass transistor switches ON. The magnitude of this drop depends on the ratio of C_1 to C_2 and the ratio of the load resistance to the pass transistor resistance. The output sequences obtained from the SPICE2 output waveform are

$$V_{out1}\text{: } (2.2 \text{ ns}, 0 \rightarrow I)(3.5 \text{ ns}, I \rightarrow 1, 5.0)$$

$$V_{out2}: (11.7 \text{ ns}, 0 \rightarrow I)(15.2 \text{ ns}, I \rightarrow 1, 3.9)$$

The reduced voltage of V_{out2} must be considered for accurate simulation of subcircuits which have V_{out2} as an input.

CHAPTER 5

SIMULATING STRONGLY CONNECTED COMPONENTS

In this chapter we discuss the use of a special windowing technique to simulate the blocks within a strongly connected component (SCC). The algorithm presented splits the entire time-interval of interest into various time-slots or windows such that all pairs of signal-transitions take place entirely within one of these windows. This is achieved by maintaining a sequential-list of intervals of transitions which is updated dynamically as the algorithm progresses. The algorithm is, in a sense, event-driven, since only those circuit-blocks that are *active* within a window are processed and the *fanouts* of the output nodes of these blocks are scheduled for processing in the future. We begin by reviewing two well-known and classical techniques, namely, the waveform relaxation method and the time-point relaxation method, that could be used to simulate the blocks within an SCC in the network. We will show that neither of these schemes is entirely suitable in our type of simulation; hence, there is a need for the event-driven windowing technique that we will present.

5.1 WAVEFORM RELAXATION vs TIME-POINT RELAXATION

Let $\Omega(N,M,\Sigma)$ be a partitioned MOS network in which the set of blocks Σ is further partitioned into its strongly-connected components $\Sigma_1,\Sigma_2,\ldots,\Sigma_\mu$. Let $[t_{start}, t_{stop}]$ denote the time-interval of simulation. Suppose Σ_i is the SCC currently scheduled for processing. If Σ_i is a simple SCC then the single block contained in it could be simulated during $[t_{start}, t_{stop}]$ by the algorithms discussed in Chapter 4. Hence, suppose that $\Sigma_i=\{\Omega_1,\Omega_2,\ldots,\Omega_p\}$, where $p\geqslant 2$ and each Ω_j is a block.

The blocks within Σ_i could then be simulated using a waveform relaxation iterative scheme WR_SIM described below. For any node $n_k \in N$ in the network, let S_k denote the most recently computed sequence of transitions or the *present-sequence* at the node and let $\hat{S}_k$ denote the previously computed sequence or the *past-sequence* at the node. Let N_i denote the list of all input and output nodes of the blocks within Σ_i. The algorithm begins by initializing the present-sequence of transitions at any node in N_i that has not been previously computed. The iterative procedure then begins by setting the past-sequence equal to the present-sequence for each node in N_i. The individual blocks within the SCC are then simulated according to a given ordering over the entire time-interval $[t_{start}, t_{stop}]$ by algorithms described in Chapter 4. In each case the present-sequences at the input nodes of a block are taken as the input sequences for simulation and the present-sequences at the output nodes of the block are updated after the simulation. The procedure EQUAL then checks for equality between the present-sequence and the past-sequence during the time-interval $[t_{start}, t_{stop}]$ at each node and returns the value 0 if they are found equal and 1 if not. Here, two sequences are considered equal if they represent the same ternary digital waveform. The iterations are carried out until both present and past sequences are found equal for each node in the SCC.

Algorithm 5.1 : Waveform Relaxation

```
procedure WR_SIM (Σi, tstart, tstop)
begin
        for each node nk ∈ Ni do
                if (Sk=Ø) then
                        INITIALIZE (Sk) ;
                end if
        end for
        repeat
                for each node nk ∈ Ni do
                        Ŝk←Sk;
                end for
                for j←1 until p do
                        SIMULATE (Ωj, tstart, tstop);
                end for
                ind←0;
                for each node nk ∈ Ni do
                        ind←EQUAL (Sk,Ŝk,tstart,tstop);
                end for
        until ind=0
end
```

We now discuss several features of the above algorithm. We first consider obtaining an *a priori* ordering on the blocks of the SCC. Given any such ordering, we define a node to be *initially relaxed* if it is an input node of a block within the SCC and its present sequence has not yet been updated in the current iteration at the time of simulating the block. In the above algorithm, the present sequence of a node gets updated only after simulating the block to which it is an output node. Hence, in the case of an initially relaxed node the blocks in its fanin list are ordered after the blocks in its fanout list. Given an ordering on the vertices of a digraph, we say that an arc is a *forward arc* if its tail vertex appears before its head vertex in the ordering ; otherwise, the arc is said to be a *feedback arc*. If we consider the vertices of the derived digraph, as defined in Chapter 3, corresponding to the blocks within the SCC Σ_i, then any ordering would result in a set of feedback arcs. Furthermore, the number of feedback arcs produced by an ordering is an upper bound on the number of initially relaxed nodes due to that ordering. Clearly, the best choice for the ordering is one which results in the least number of initially relaxed nodes since this would speed up the convergence of the above algorithm. However, this corresponds to finding an ordering that results in the minimum number of feedback arcs, which is an NP-Complete problem [56-58]. Therefore, the choice of the *a priori* ordering affects the speed of convergence of the above algorithm and finding the best ordering, in this respect, turns out to be a difficult problem from the computational complexity point of view. This is one of the drawbacks of the waveform relaxation scheme.

Another aspect that needs to be considered is that the number of iterations turns out to be proportional to the number of transitions at the various circuit nodes in certain circuits such as the ring oscillator. This is also one of the major drawbacks in the waveform relaxation method WRM [10–12]. Finally, this scheme requires storing two sequences of transitions for the entire time-interval $[t_{start}, t_{stop}]$ at each node which could be a considerable amount of computer storage for large SCCs. In spite of all these drawbacks, this scheme could still be used in our type of switch-level simulation since it is easy to implement and is compatible with our delay operations.

An alternative approach is to use the time-point relaxation method for the simulation of the entire partitioned network $\Omega(N,M,\Sigma)$. In this approach there is no need to handle blocks within an SCC in a special way since the scheme is *event-driven*, as discussed in Chapter 2, and is used in several digital simulators [13,17,20,23,28,29]. In order to use this approach in our type of simulation, we could define an *event* as a transition $(t_i, x \rightarrow y)$ occurring at time t_i. A time-queue (TQ) is used to maintain a list of events occurring at different instants of time. If a transition $(t_i, x \rightarrow y)$ occurs at some node n_j in the network, then all the blocks in the fanout list of n_j are processed at time t_i. If on processing a block at t_i, a transition is observed at an output node of the block, then this is defined as a new event, and is scheduled to occur at time $t'_i > t_i$. Thus, $t'_i - t_i > 0$ is a positive delay in propagating an event occurring at an input node of a block to an output node of the block. It is this feature that makes the use of the time-point relaxation approach particularly attractive for processing blocks within feedback loops.

In our type of switch-level simulation, the emphasis is on generating accurate timing estimates which is possible by using the delay operations described in Chapter 4. However, the delay operator can only operate on a pair of complete transitions; therefore, events can be propagated through a block only in pairs. Consider an example of an inverter with a sequence $(t_1,0 \rightarrow I)$, $(t_2,I \rightarrow 1)$ at its input node causing a sequence $(t'_1,1 \rightarrow I)$, $(t'_2,I \rightarrow 0)$ at its output node. In order to use the time-point relaxation scheme, we would have to be able to compute the value of t'_1 only with the knowledge of the input event $(t_1,0 \rightarrow I)$. This is however impossible, since the delay operator needs to know the values of both t_1 and t_2 before it can compute t'_1 and t'_2. Furthermore, it is possible to have a situation where $t'_2 < t_2$, which means that the input event $(t_2,I \rightarrow 1)$ causes the output event $(t'_2,I \rightarrow 0)$ at an earlier time, thus violating the basic assumption that one only advances in time in the TQ and never has to backtrack. Therefore, the time-point relaxation method, as such, is not suitable for our type of simulation.

5.2 DYNAMIC WINDOWING

In the previous section we discussed two relaxation methods to simulate the blocks in a network. The first method, namely, the waveform relaxation method, could be used in our type of simulation since it is compatible with the delay operations, but suffers from several drawbacks in the case of blocks within strongly connected components. The second method, namely, the time-point relaxation method, is used in several digital simulators, mainly because blocks within strongly-connected components do not pose any special problems, but it is found to be incompatible with the delay operations described in Chapter 4; hence, it cannot be used, as such, in our type of simulation. In this section we describe a new scheme to handle blocks within an SCC which overcomes most of the above drawbacks in the waveform relaxation method by incorporating some of the ideas of the time-point relaxation method. The main idea is to use the so-called *windowing* technique in the waveform-relaxation procedure, as suggested in [11,12], wherein it is shown that the number of iterations is exponentially proportional to the size of the time-interval of analysis. This suggests dividing the entire time-interval of interest into many time-slots or windows so that waveform relaxation can be performed within each window. These waveforms generate initial conditions for the next window and, so on. If all the windows have the same size, then there exists an optimum number of windows which minimizes the total number of iterations; hence, the total CPU time for analysis, as shown in [11].

The choice of windows, however, is very crucial in our type of switch-level simulation since, in order to obtain good timing through the delay-operator, the window boundaries must be chosen such that all the intervals during which the signal at a node in the SCC is is switching (i.e., in the **SWLO** or **SWHI** states) must be placed entirely within a window. This appears to be a no-win situation since deciding on the placement of windows seems to require *a priori* knowledge of the digital waveform (or sequences of transitions) at each circuit-node within the SCC. Here we describe a successful solution to this problem by using a sequential-list of time-intervals which is dynamically updated as the algorithm progresses. In addition, the new scheme is event-driven, and therefore, requires no *a priori* ordering of blocks within a SCC. Before going into the description of the algorithm, a few definitions and notations are needed.

Consider an SCC Σ_i consisting of a set of blocks $\Omega_1, \Omega_2, \ldots, \Omega_p$. Let EXT_i denote those circuit nodes in the blocks within the SCC for which the node-sequences have already been computed. For each circuit-node n_k in the SCC, let $FO(n_k) = FOUT(n_k) \cap \Sigma_i$ denote the set of blocks within Σ_i for which n_k is an input node.

Definition : A *transition-interval* for a node is the time-interval during which the signal at the node is in the **SWLO**, **SWHI**, or **MID** state. Associated with each transition-interval **I** for a node n_k is a fanout list of blocks, denoted by $F(I)$, which is initially set to $FO(n_k)$. Let $a(I)$ and $b(I)$ denote the initial and final times of the transition-interval **I**.

Let I_1 and I_2 be any two transition-intervals. We say that $I_1 < I_2$ if and only if $b(I_1) < a(I_2)$. If $I_1 \cap I_2 \neq \emptyset$, then we say that I_1 and I_2 are *incomparable*. This defines a partial order "<" on a set of intervals. Let $L = \{I_1, I_2, \ldots, I_q\}$ be a sequential list of intervals. We say that **L** is an *ordered-list* if $I_1 < I_2 < \cdots < I_q$. We say that an interval **I** is *contained* in **L** if $I \subseteq I_j$ for some $I_j \in L$. Given any interval **I** and an ordered list of intervals **L** the following procedure returns an updated ordered list $\hat{L}$ containing the interval **I**.

```
procedure INCLUDE(I,L)
begin
        L̂←∅;
        η←0;
        for j← 1 until q do
                if I_j<I then
                        L̂←L̂ ∪ I_j;
                else if I ∩ I_j≠∅ then
                        I←I ∪ I_j;
                        F(I)←F(I) ∪ F(I_j);
                else if I<I_j then
                        if η=0 then
                                L̂←L̂ ∪ I;
                        end if
                        η←1;
                        L̂←L̂ ∪ I_j;
                end if
        end for
        if η=0
                L̂←L̂ ∪ I;
        return L̂;
end
```

The algorithm for the new dynamic windowing technique can now be described as follows. The ordered set L is initialized to the empty set. Every transition-interval at each node in EXT_i is included in L. The set L is altered, dynamically, as the algorithm progresses. At any stage, we have a partition of the entire time-interval $[t_{start}, t_{stop}]$ into windows by taking the final times of the disjoint intervals in L as the boundaries of the windows. The set L plays the role of the time-queue (TQ) used in the time-point relaxation method. Here events take place over transition-intervals rather than time-instants. If a transition-interval at an input node of a block causes a transition-interval at an output node of the block, then the end-points of the new interval are computed by the delay operator described in Chapter 4.

The iterative loop in Algorithm 5.2 begins by choosing the first interval in L as the window of interest. The blocks in its fanout list are then simulated only for the duration of the present window until the list is empty. Each time a block gets simulated, a transition-interval in an output node is included in L if and only if one of the following two conditions are satisfied:

a) All transitions in the output node occur after the present window, I.

b) The transitions at the output node, occurring during the present window after processing the block, are different from those before processing the block.

After the fanout list for the present window is empty the interval is deleted from L and the whole process is repeated until L is empty. The procedure **EXTRACT**(S_k,t_{ini},t_{fin}) returns a sequence of only those transitions in S_k occurring between t_{ini} and t_{fin}.

Consider the execution of Algorithm 5.2 on an SCC Σ_i. After the initialization of L by including the transition-intervals of the nodes in EXT_i, it could get updated by the inclusion of the transition-intervals at the output nodes of the block that has been just simulated. This could alter either the endpoint, t_{fin}, of the present window, or could append a set of blocks to the existing fanout list F(I) of the present window. If

Algorithm 5.2 : Dynamic windowing

```
procedure DYNAMIC_WINDOW (Σi, tstart, tstop)
begin
        L←Ø;
        for each circuit-node nk∈EXTi do
                for each transition-interval Ij of nk do
                        L←INCLUDE (Ij,L)
                end for
        end for
        tfin←tstart;
        while L is not empty do
                I← first interval in L;
                tini←tfin;
                while F(I) is not empty do
                        tfin←b(I);
                        Ωr← first block in F(I);
                        delete the first block from F(I);
                        for each node nk∈OUT(Ωr) do
                                Ŝk←EXTRACT(Sk,tini,tfin);
                        end for
                        SIMULATE (Ωr, tini, tfin);
                        for each output node nk of Ωr do
                                S'k←EXTRACT(Sk,tini,tfin);
                                for each transition-interval Im of nk do
                                        if I<Im then
                                                L←INCLUDE(Im,L);
                                        else if S'k≠Ŝk then
                                                L←INCLUDE(Im,L);
                                        end if
                                end for
                        end for
                end while
                delete the first interval from L;
        end while
end
```

the latter situation continues, it is possible that a block could reappear in the fanout list of the present window, after it has been deleted before, and is, therefore, resimulated during the present window. We say that an SCC is *well-behaved* if, during the execution of Algorithm 5.2, none of its blocks is ever resimulated during the same window.

Thus, in a well-behaved SCC the delay characteristics of the various blocks are such that one does not have to perform any iterations at all. If, however, the SCC is not well-behaved, then the algorithm extends the fanout list of the present window, and resimulates the active blocks until convergence is achieved for the duration of the present window. This is equivalent to performing waveform relaxation iterations within the present window. It is possible to conjure up an SCC for which the initial window gets continually extended until it becomes the entire time-interval. In this case using Algorithm 5.2 becomes equivalent to using Algorithm 5.1. However, such a situation is of theoretical interest only, and probably never occurs in practical circuits. Thus, in the worst case, the new dynamic windowing techniques performs at least as well as the waveform relaxation method. In fact, the SCCs in several practical circuits considered were all well-behaved, in which case Algorithm 5.2 performs much better than the waveform relaxation Algorithm 5.1 in all respects. First, there is no need to place the blocks of the SCC in any particular order, since the procedure is *event-driven*, i.e., only those blocks that are *active* in a window are processed during that window. Second, no iterations are performed in the case of a well-behaved SCC, thereby saving considerable amounts of computation time, and finally, the active blocks are processed only during a window (and not for the entire time-interval), thus causing a reduction in both computation time and memory space required to store the sequences of transitions.

CHAPTER 6

PERFORMANCE OF IDSIM2

In this chapter we discuss the performance of IDSIM2, a switch-level timing simulator for MOS circuits using the algorithms described in the earlier chapters. IDSIM2 has been implemented in the C programming language and runs on a SUN 3/160 workstation using the UNIX operating system. The MOS network is described to IDSIM2 in a SPICE2 input format.

We evaluate the performance of IDSIM2 based on its computational speed (complexity) and the accuracy of its switch-level timing estimates. We first evaluate the computational speed by considering several examples. The first example is a combinatorial CMOS NAND gate implementation of a one-bit full adder circuit, shown in Figure 6.1, which was cascaded to produce full-adders from one to eight bits. Table 6.1 shows the rate of growth of CPU-time versus the number of transistors. The CPU-time shown in the table is the time for dc and transient analysis for IDSIM2 and SPICE2G.5. This table shows that the simulation time taken by IDSIM2 is fairly linear with circuit size and ranges from 600 to 2400 times faster than SPICE2G.5.

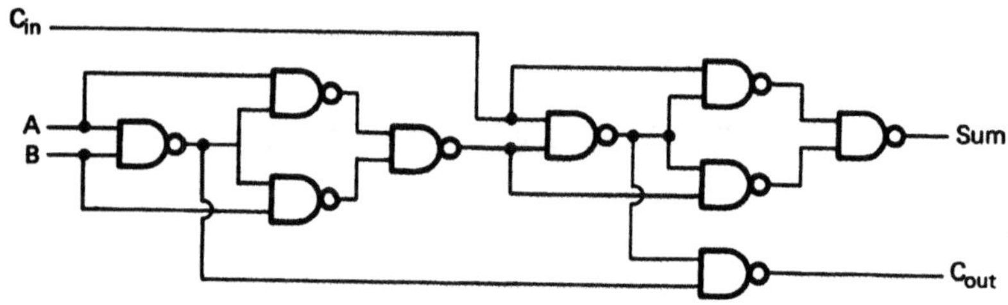

Figure 6.1 : A CMOS one bit adder

Table 6.1: The growth-rate of CPU-time of SPICE2 and IDSIM2

Adder Bits	Number of Transistors	CPU - seconds SPICE2G.5	IDSIM2
1	36	133.79	0.22
2	72	1112.16	0.96
4	144	2181.62	1.56
6	216	5102.97	2.46
8	288	7747.91	3.16

The outputs of bits 0, 2, 4, and 5 and the carry of the six bit adder are shown in Figure 6.2. Throughout this chapter we will represent analog waveforms produced by SPICE2G.5 with solid lines and the ternary digital waveforms produced by IDSIM2 with dotted lines.

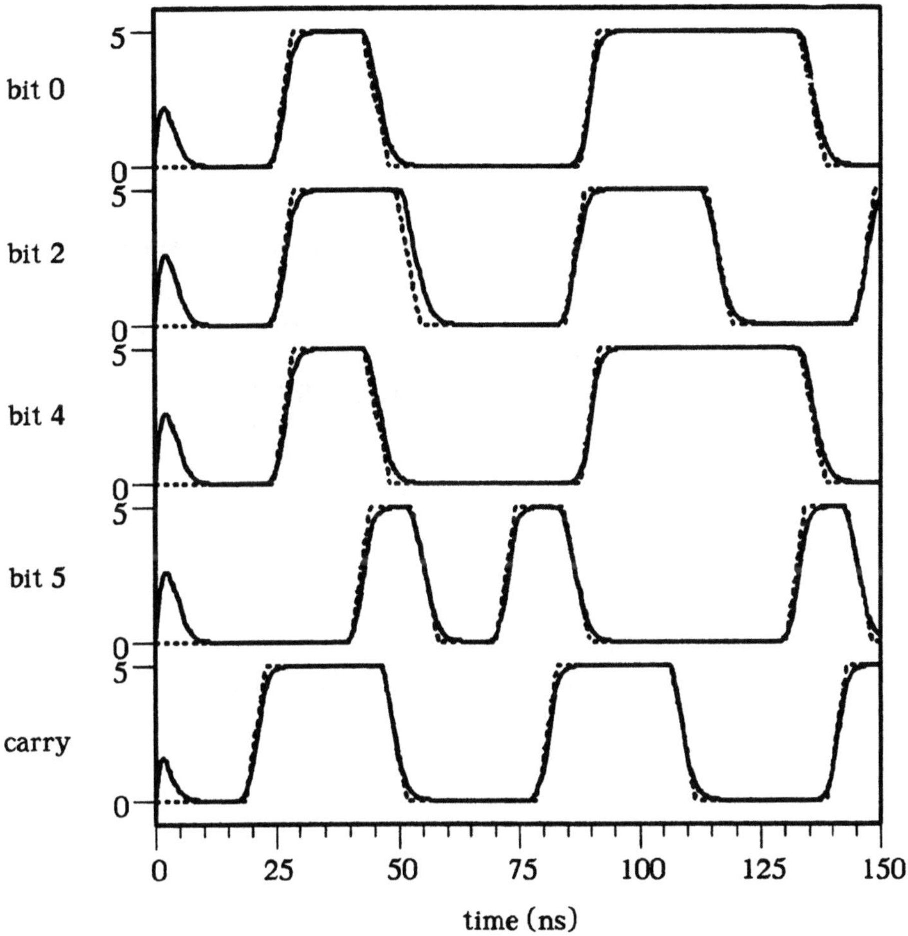

Figure 6.2 : Waveforms for a six bit CMOS adder

The second example is a chain of identical CMOS inverters. Figure 6.3 shows a chain of five inverters. Table 6.2 below gives the CPU-times taken by both IDSIM2 and SPICE2G.5 as a function of the chain length. The waveforms at the output of every tenth inverter in a 50-inverter chain produced by both IDSIM2 and SPICE2G.5 are shown in Figure 6.4.

Table 6.2: The CPU-times taken on a chain of inverters

Number of Inverters	CPU - seconds	
	SPICE2G.5	IDSIM2
5	16.23	0.10
10	32.33	0.20
20	77.17	0.44
30	135.42	0.60
40	212.18	0.84
50	299.04	1.04

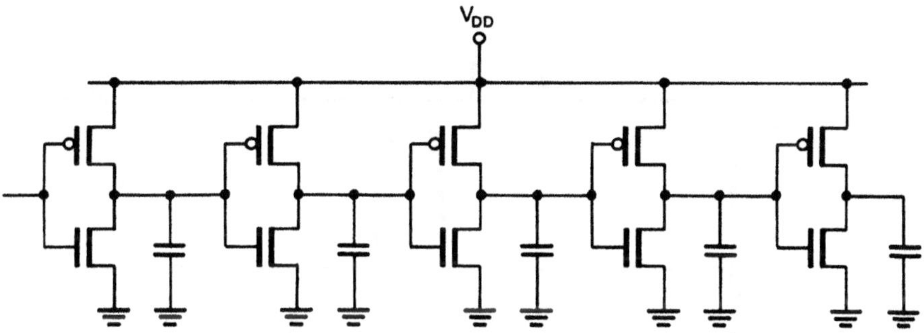

Figure 6.3 : A five-stage inverter chain

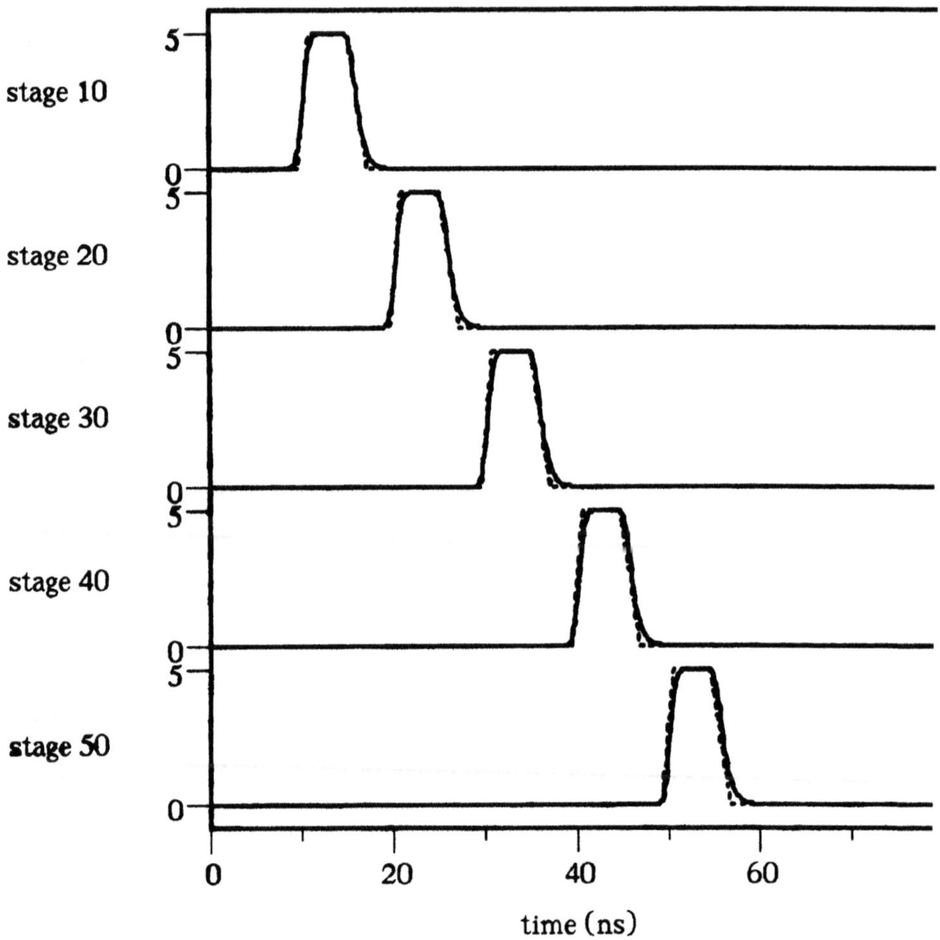

Figure 6.4 : Waveforms for a 50-inverter chain

From both of the examples considered above, it can be concluded that the CPU-time taken by IDSIM2 grows linearly with circuit size and is at least two orders of magnitude faster than SPICE2G.5.

We now consider several examples of circuits simulated using IDSIM2 to illustrate its timing accuracy. A three-stage CMOS ring oscillator is shown in Figure 6.5. This may be a difficult type of circuit to simulate in switch-level timing simulation due to its analog behavior. The waveforms for this circuit are shown in Figure 6.6.

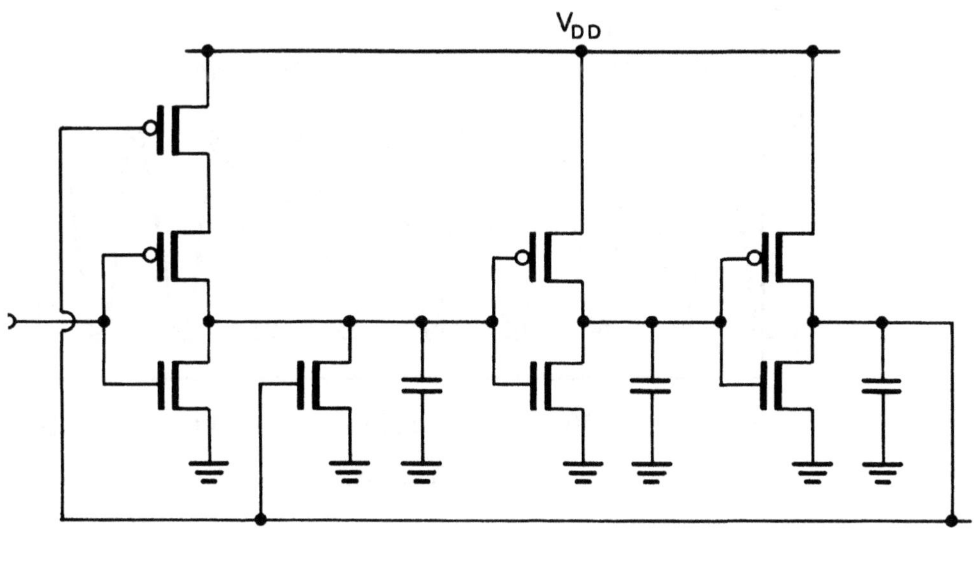

Figure 6.5 : A CMOS ring oscillator

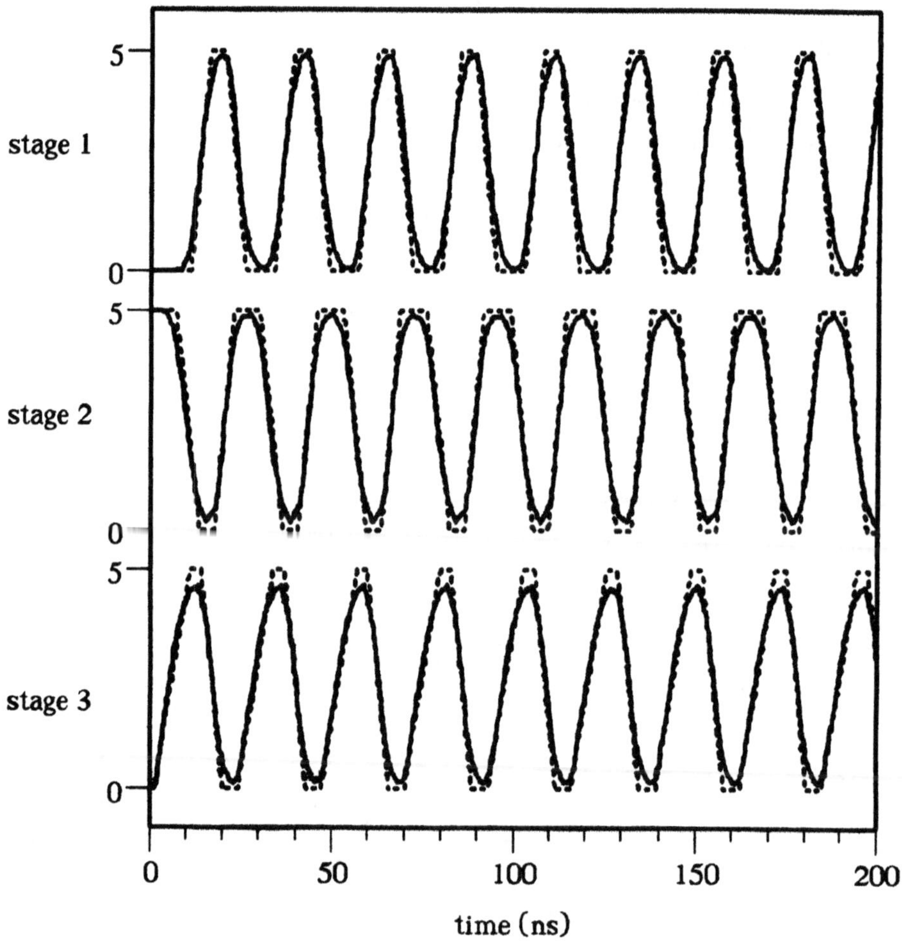

Figure 6.6 : Waveforms for a CMOS ring oscillator

An NMOS tally circuit implemented using pass transistors is shown in Figure 6.7. An important aspect of the simulation of this circuit is the voltage drop across the pass transistors. The waveforms for this circuit are shown in Figure 6.8. Since the simulator waveforms have only three logic states, partial transitions (glitches) are modeled using a temporary transition into the I state. A nice feature of IDSIM2 is that only glitches that cross one of the logic thresholds (i.e., V_{TL} or V_{TH}) appear as partial transitions in the final waveform. Minor glitches that do not cross either of the logic thresholds are ignored. This circuit also illustrates the accuracy of timing estimates for a chain of pass transistors.

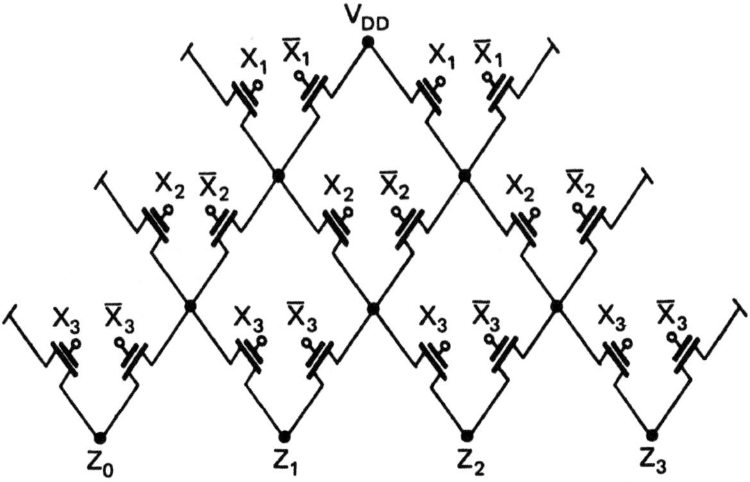

Figure 6.7 : An NMOS tally circuit

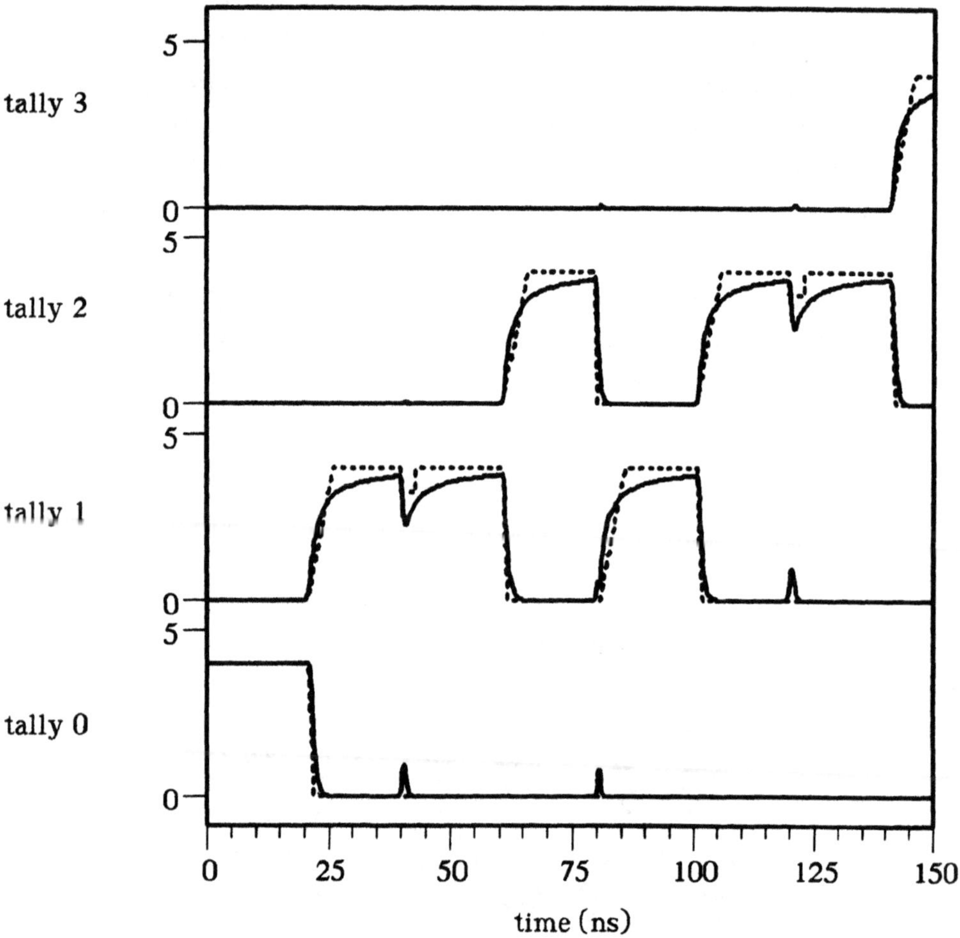

Figure 6.8 : Waveforms for an NMOS tally circuit

The simulation of NMOS circuits serves as a means to test the accuracy of the calculation of the dc output voltage described in Section 4.5 of Chapter 4 since the load transistor in such circuits is always in the ON state. An NMOS bridge network is shown in Figure 6.9 and its output waveform is shown in Figure 6.10.

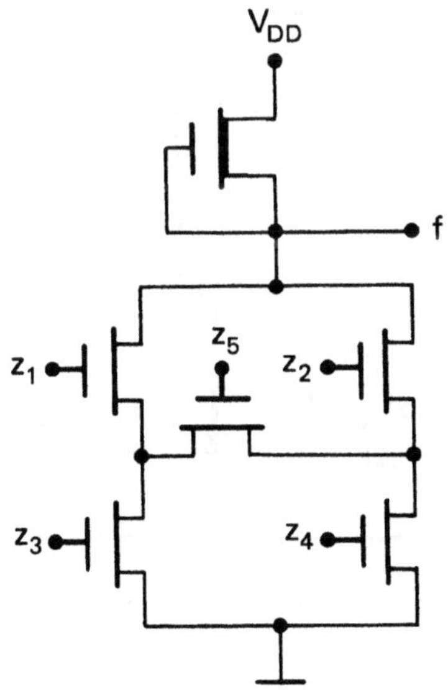

Figure 6.9 : An NMOS bridge network

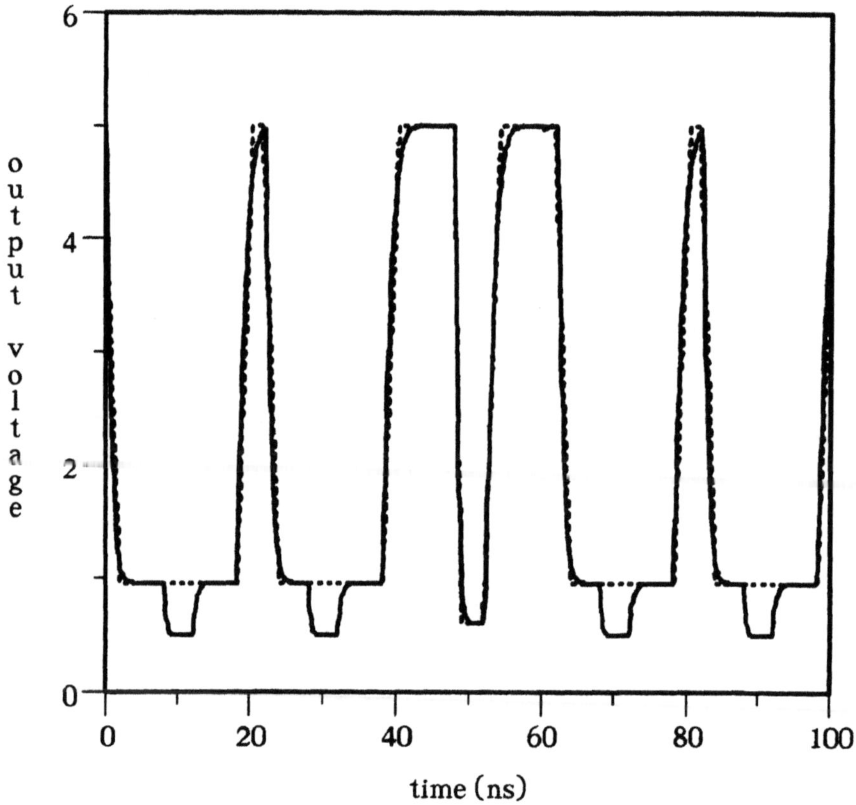

Figure 6.10 : Output Waveform for an NMOS bridge network

A possible test of the the accuracy of voltage calculations in the equations of Section 4.4 of Chapter 4 is to simulate an inverter driving a pass transistor, where the inverter and the pass transistor control signal are pulsed at different frequencies. A CMOS inverter driving a pass transistor is shown in Figure 6.11. The waveforms for this circuit are shown in Figure 6.12. The difficulty with this circuit is to accurately calculate the voltage at the output nodes. Tabular approaches to delay cannot determine the voltage on the output node when the pass transistors are switched OFF, and thus, cannot find accurate waveforms for a circuit with this type of behavior. Since IDSIM2 performs calculations of the output node voltage, the voltage is saved when the pass transistor is switched OFF. Thus, the calculation of the output node voltage changes continues when the pass transistor switches ON again.

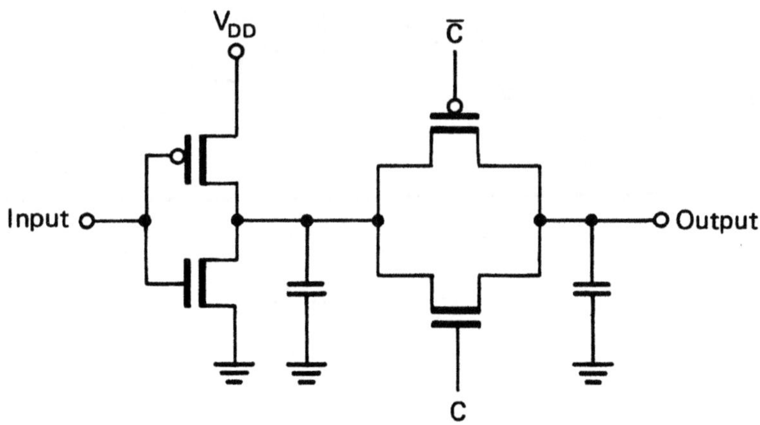

Figure 6.11 : A CMOS inverter driving a pass transistor

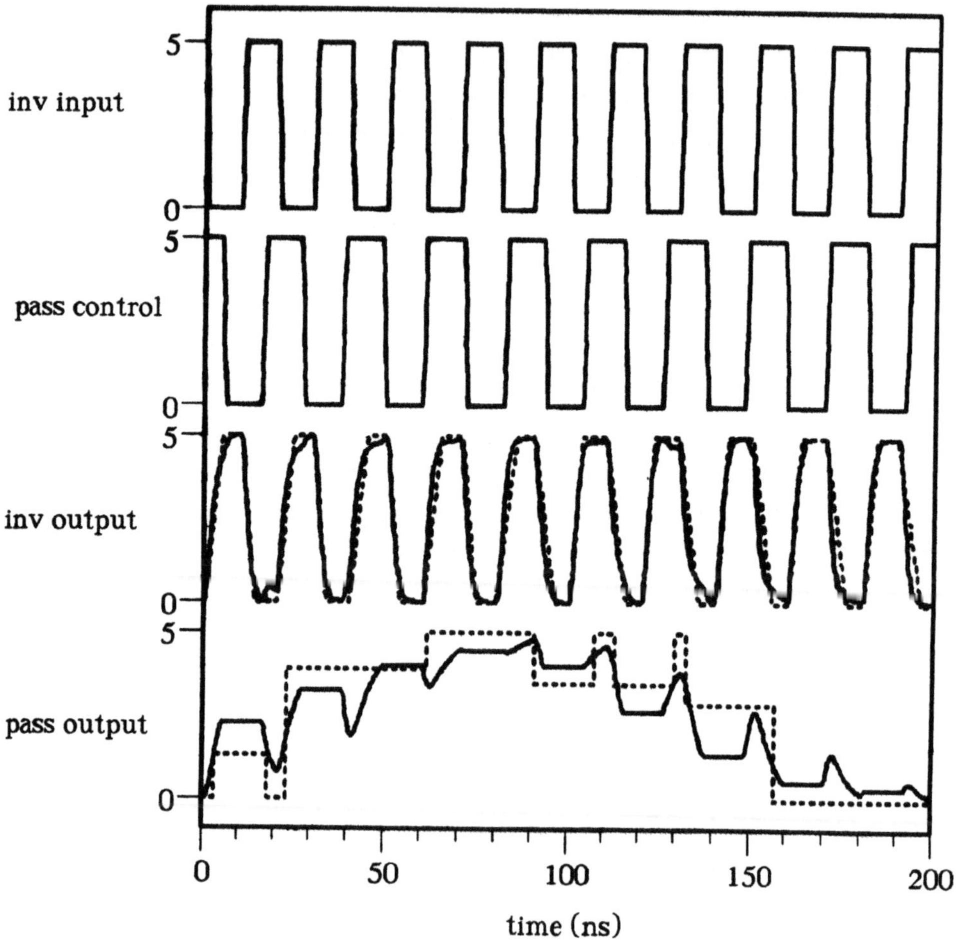

Figure 6.12 : The waveforms for an inverter driving a pass transistor

In Table 6.3 we provide a list of some of the circuits that have been simulated using IDSIM2 thus far, along with the number of transistors, and the CPU-time taken by IDSIM2 and SPICE2G.5.

Table 6.3: Circuits simulated by IDSIM2

Circuit	Number of transistors	CPU-time	
		SPICE2G.5	IDSIM2
NMOS bridge network	11	177.43	0.22
NMOS ring oscillator	31	68.08	0.34
NMOS inverter chain	100	463.50	2.08
CMOS ring oscillator	12	180.13	0.88
CMOS inverter chain	100	945.34	2.08
CMOS edge detection	57	405.42	0.78
CMOS adder with parasitics	44	487.53	2.26
CMOS feedback loops	132	2377.63	10.98
CMOS logic gates	1434	-	2.60

The waveforms in Figures 6.2, 6.4, 6.6, 6.8, 6.10, and 6.12 indicate the excellent accuracy of the switch-level timing estimates generated by IDSIM2. For purely digital circuits the timing accuracy is within 5% of that of SPICE2G.5. Furthermore, IDSIM2 performs simulations at speeds of two to three orders of magnitude faster than SPICE2G.5.

REFERENCES

[1] L. W. Nagel, "SPICE2: A Computer Program to Simulate Semiconductor Circuits," Electronics Research Laboratory Report #ERL-M520, University of California, Berkeley, May 1975.

[2] W. T. Weeks, A. J. Jimenez, G. W. Mahoney, D. Mehta, H. Qassemzadeh, and T. R. Scott, "Algorithms for ASTAP - A Network Analysis Program," *IEEE Transactions on Circuit Theory*, vol. CT-20, pp. 628-634, November 1973.

[3] P. Yang, I. N. Hajj, and T. N. Trick, "SLATE: A Circuit Simulation Program with Latency Exploitation and Node Tearing," *Proceedings of the IEEE International Conference on Circuits and Computers*, pp. 353-355, October 1980.

[4] N. B. G. Rabbat, A. L. Sangiovanni-Vincentelli, and H. Y. Hsieh, "A Multi-level Newton Algorithm with Macromodelling and Latency for the Analysis of Large-Scale Nonlinear Circuits in the Time Domain," *IEEE Transactions on Circuits and Systems*, vol. CAS-26, pp. 733-741, September 1979.

[5] B. R. Chawla, H. K. Gummel, and P. Kozak, "MOTIS - An MOS Timing Simulator," *IEEE Transactions on Circuits and Systems*, vol. CAS-22, pp. 901-910, December 1975.

[6] S. P. Fan, M. Y. Hsueh, A. R. Newton, and D. O. Pederson, "MOTIS-C: A New Circuit Simulator for MOS LSI Circuits," *Proceedings of the IEEE International Symposium on Circuits and Systems*, Phoenix, pp. 700-703, April 1977.

[7] C. F. Chen, C. Y. Lo, H. N. Nham, and P. Subramaniam, "The Second Generation MOTIS Mixed-Mode Simulator," *Proceedings of the 21st Design Automation Conference*, Albuquerque, New Mexico, pp. 10-17, June 1984.

[8] Y. P. Wei, I. N. Hajj, and T. N. Trick, "A Prediction-Relaxation Based Simulator for MOS Circuits," *Proceedings of the IEEE International Conference on Circuits and Computers*, New York, September 1982.

[9] E. Lelarasmee, "The Waveform Relaxation Method for the Time Domain Analysis of Large Scale Non-linear Dynamical Systems," Ph.D. Dissertation, University of California, Berkeley, 1981.

[10] E. Lelarasmee, A. E. Ruehli, and A. L. Sangiovanni-Vincentelli, "The Waveform Relaxation Method for the Time Domain Analysis of Large Scale Integrated Circuits," *IEEE Transactions on Computer-Aided Design*, vol. CAD-1, no. 3, pp. 131-145, July 1982.

[11] J. White and A. L. Sangiovanni-Vincentelli, "RELAX2 : A Modified Waveform Relaxation Approach to the Simulation of MOS Digital Circuits," *Proceedings of the IEEE International Symposium on Circuits and Systems*, California, pp. 756-759, May 1983.

[12] J. White and A. L. Sangiovanni-Vincentelli, "RELAX2.1 : A Waveform Relaxation based Circuit Simulation Program," *Proceedings of the IEEE Custom Integrated Circuits Conference*, Rochester, New York, pp. 232-236, May 1984.

[13] A. R. Newton, "The Simulation of Large-Scale Integrated Circuits," *IEEE Transactions on Circuits and Systems*, vol. CAS-26, pp. 741-749, September 1979.

[14] G. Arnout and H. De Man, "The use of Threshold Functions and Boolean-Controlled Network Elements for Macromodelling of LSI Circuits," *IEEE Journal of Solid State Circuits*, vol. SC-13, pp. 326-332, June 1978.

[15] K. Sakallah and S. W. Director, "An Activity Directed Circuit Simulation Algorithm," *Proceedings of the IEEE International Conference on Circuits and Computers*, New York, pp. 356-360, October 1980.

[16] G. R. Case, "SALOGS - A CDC 6600 Program to Simulate Digital Logic Networks, vol. 1 - User's Manual," *Sandia Laboratory Report SAND 74-0441*, 1975.

[17] S. A. Szygenda, "TEGAS2 - Anatomy of a General Purpose Test Generation and Simulation System for Digital Logic," *Proceedings of the Ninth ACM Design Automation Workshop*, June 1972.

[18] J. Jephson, R. McQuarrie, and R. Vogelsberg, "A Three-Value Computer Design Verification System," *IBM Systems Journal*, vol. 8, no. 3, pp. 178-188, 1969.

[19] S. Seshu and D. N. Freeman, "The Diagnosis of Asynchronous Sequential Switching Systems," *IRE Transactions on Electronic Computing*, vol. EC-11, No. 4, pp. 459-465, August 1962.

[20] E. G. Ulrich, "Exclusive Simulation of Activity in Digital Networks," *Communication of the ACM*, vol. 12, no. 2, pp. 102-110, February 1969.

[21] S. A. Szygenda and E. W. Thompson, "Modelling and Digital Simulation for Design Verification and Diagnosis," *IEEE Transactions on Computers*, vol. C-25, pp. 1242–1253, December 1976.

[22] P. Wilcox, "Digital Logic Simulation at the Gate and Functional Level," *Proceedings of the IEEE Design Automation Conference*, New York, pp. 242–248, 1979.

[23] R. E. Bryant, "An Algorithm for MOS Logic Simulation," *LAMBDA (now VLSI) magazine*, vol. 1, no. 3, pp. 46–53, 1980.

[24] R. E. Bryant, "A Switch-level Simulation Model for Integrated Logic Circuits," Ph.D. dissertation, MIT/LCS/TR-259, Massachusetts Institute of Technology, Cambridge, March 1981.

[25] R. E. Bryant, "A Switch-level Model and Simulator for MOS Digital Circuits," *IEEE Transactions on Computers*, vol. C-33, no. 2, pp. 160–177, February 1984.

[26] R. H. Byrd, G. D. Hachtel, M. R. Lightner, and M. H. Heydemann, "Switch Level Simulation : Models, Theory and Algorithms," Advances in Computer-Aided Engineering Design, A. L. Sangiovanni-Vincentelli, Editor. Jai Press, 1985.

[27] V. Ramachandran, "An Improved Switch-level Simulator for MOS Circuits," *Proceedings of the 20th Design Automation Conference*, Miami Beach, Florida, pp. 293–299, June 1983.

[28] I. N. Hajj and D. G. Saab, "Symbolic Logic Simulation of MOS Circuits," *Proceedings of the IEEE International Symposium on Circuits and Systems*, Newport Beach, California, pp. 246-249, May 1983.

[29] C. J. Terman, "RSIM - A Logic-Level Timing Simulator," *Proceedings of the IEEE International Conference on Computer Design*, New York, pp. 437-440, November 1983.

[30] C. J. Terman, "Simulation Tools for Digital LSI Design," Ph.D. dissertation, Department of Electrical Engineering and Computer Science, Massachusetts Institute of Technology, Cambridge, 1983.

[31] V. B. Rao, T. N. Trick, and I. N. Hajj, "A Table-Driven Delay-Operator Approach to Timing Simulation of MOS VLSI Circuits," *Proceedings of the IEEE International Conference on Computer Design*, New York, pp. 445-448, November 1983.

[32] Vasant B. Rao, "Switch-Level Timing Simulation of MOS VLSI Circuits," Ph.D. dissertation, Department of Electrical and Computer Engineering, University of Illinois, Urbana, January 1985.

[33] D. Overhauser, I. N. Hajj, and V. B. Rao, "Switch-Level Timing Analysis of VLSI MOS Circuits," *Proceedings of the IEEE International Symposium on Circuits and Systems*, San Jose, California, pp. 761-764, May 1986.

[34] F-P Lai, V. B. Rao, and T. N. Trick, "JADE: A Hierarchical Switch Level Timing Simulator," *Proceedings of the IEEE International Symposium on Circuits and Systems*, Philadelphia, pp. 592-595, May 1987.

[35] H. N. Nham and A. K. Bose, "A Multiple Delay Simulator for MOS LSI Circuits," *Proceedings of the 17th Design Automation Conference*, pp. 610-617, June 1980.

[36] Vasant B. Rao, "Algorithms for a Multiple-Delay Simulator," M. S. thesis, Department of Electrical Engineering, University of Illinois, Urbana, March 1982.

[37] K. Okazaki, T. Moriya, and T. Yahara, "A Multiple Media Delay Simulator for MOS LSI Circuits," *Proceedings of the 20th Design Automation Conference*, pp. 279-285, June 1983.

[38] T. Tokuda, K. Okazaki, K. Sakashita, I. Ohkura, and T. Enomoto, "Delay-Time Modeling for ED MOS Logic LSI," *IEEE Transactions on Computer-Aided Design*, vol. CAD-2, no. 3, pp. 129-134, July 1983.

[39] T. J. Shafer, "A Transistor-Level Logic-With-Timing Simulator for MOS Circuits," *Proceedings of the 22nd Design Automation Conference*, pp. 762-765, 1985.

[40] M. Yoeli and S. Rinon, "Application of Ternary Algebra to the Study of Static Hazards," *Journal of the ACM*, vol. 11, no. 1, pp. 84-97, January 1964.

[41] E. B. Eichelberger, "Hazard Detection in Combinatorial and Sequential Switching Circuits," *IBM Journal of Research and Development*, vol. 9, pp. 90-99, March 1965.

[42] V. B. Rao, T. N. Trick, and M. R. Lightner, "Hazard Detection in a Multiple Delay Logic Simulator," *Proceedings of the IEEE International Symposium on Circuits and Systems*, Rome, Italy, pp. 72-75, May 1982.

[43] R. E. Bryant, "Race Detection in MOS Circuits by Ternary Simulation," *VLSI 83*, F. Anceau, Editor. North-Holland Publishing Company, August 1983.

[44] V. Ramachandran, "A Linear Time Algorithm for Race Detection in Transistor Switch-level Circuits," *Proceedings of the IEEE International Conference on Computer Design*, New York, pp. 345-348, November 1983.

[45] W. C. Elmore, "The Transient Response of Damped Linear Networks with Particular Regard to Wideband Amplifiers," *Journal of Applied Physics*, vol. 19, pp. 55-64, January 1948.

[46] J. Rubinstein, P. Penfield Jr., and M. A. Horowitz, "Signal Delays in RC Tree Networks," *IEEE Transactions on Computer-Aided Design*, vol. CAD-2, no. 3, pp. 202-211, July 1983.

[47] M. A. Horowitz, "Timing Models for MOS Pass Networks," *Proceedings of the IEEE International Symposium on Circuits and Systems*, Newport Beach, California, pp. 198-201, May 1983.

[48] J. L. Wyatt, and Q-J Yu, "Signal Delay in RC Meshes, Trees and Lines," *Proceedings of the IEEE International Conference on Computer Aided Design*, Santa Clara, pp. 15-17, November 1984.

[49] C. A. Zukowski, "Relaxing Bounds for Linear RC Mesh Circuits," *IEEE Transactions on Computer-Aided Design*, vol. CAD-5, no. 2, pp. 305-311, April 1986.

[50] J. L. Wyatt, "Waveform Bounding for Fast Timing Analysis of MOS VLSI Circuits," *Proceedings of the IEEE International Symposium on Circuits and Systems*, Newport Beach, California, pp. 760-761, May 1983.

[51] J. L. Wyatt Jr., C. A. Zukowski, L. A. Glasser, P. Bassett, and P. Penfield Jr., "The Waveform Bounding Approach to Timing Analysis of Digital MOS IC's," *Proceedings of the IEEE International Conference on Computer Aided Design*, Santa Clara, pp. 392-395, November 1983.

[52] N. P. Jouppi, "TV: An NMOS Timing Analyzer," *Proceedings of the Third Caltech Conference on VLSI*, pp. 71-85, 1983.

[53] J. K. Ousterhout, "A Switch-Level Timing Verifier for Digital MOS VLSI," *IEEE Transactions on Computer-Aided Design*, vol. CAD-4, no. 3, pp. 336-349, July 1985.

[54] R. Tarjan, "Depth-First Search and Linear Graph Algorithms," *SIAM Journal of Computing*, vol. 1, no. 2, pp. 146-160, June 1972.

[55] J. A. Bondy and U. S. R. Murty, *Graph Theory with Applications*, North-Holland, 1982.

[56] A. V. Aho, J. E. Hopcroft, and J. D. Ullman, *The Design and Analysis of Computer Algorithms*, Addison-Wesley, 1974.

[57] M. R. Garey and D. S. Johnson, *Computers and Intractability : A Guide to the Theory of NP-Completeness*, W. H. Freeman and Company, 1979.

[58] S. Even, *Graph Algorithms*, Computer Science Press, 1979.

[59] E. M. Reingold, J. Nievergelt, and N. Deo, *Combinatorial Algorithms : Theory and Practice*, Englewood Cliffs, New Jersey, Prentice-Hall Inc., 1977.

[60] A. V. Vasquez, S. W. Director, and K. A. Sakallah, "PRIMO: A VLSI Circuit Partitioner for Simulation Applications," *Proceedings of the IEEE International Symposium on Circuits and Systems*, Kyoto, Japan, pp. 1075-1078, June 1985.

[61] A. Gupta, "ACE : A Circuit Extractor," *Proceedings of the ACM-IEEE 20th Design Automation Conference*," Miami Beach, Florida, pp. 721-725, June 1983.

[62] C. M. Baker, "Artwork Analysis Tools for VLSI Circuits," M. S. thesis, Department of Electrical Engineering and Computer Science, Massachusetts Institute of Technology, Cambridge, June 1980.

[63] M. Y. Tsai, "Pass Transistor Networks in MOS Technology: Synthesis, Performance, and Testing," *Proceedings of the IEEE International Symposium on Circuits and Systems*, Newport Beach, California, pp. 509-512, May 1983.

[64] R. E. Bryant, "A Survey of Switch-Level Algorithms," *IEEE Design and Test of Computers*, vol. 4, no. 4, pp. 26-40, August 1987.

[65] J. P. Hayes, "A Unified Switching Theory with Applications to VLSI Design," *Proceedings of the IEEE*, pp. 1140-1141, October 1982.

[66] I. N. Hajj, "A Path Algebra for Switch-Level Simulation," *Proceedings of the IEEE International Conference on Computer-Aided Design*, Santa Clara, California, November 1985.

[67] R. E. Bryant, "Algorithmic Aspects of Symbolic Switch Network Analysis," *IEEE Transactions on Computer-Aided Design*, vol. CAD-6, no. 4, pp. 618-639, July 1987.

[68] R. E. Bryant, "Boolean Analysis of MOS Circuits," *IEEE Transactions on Computer-Aided Design*, vol. CAD-6, no. 4, pp. 634-649, July 1987.

[69] E. Cerny and J. Gecsei, "Simulation of MOS Circuits by Decision Diagrams," *IEEE Transactions on Computer-Aided Design*, pp. 685-693, October 1985.

[70] D. S. Reeves and M. J. Itwin, "Fast Methods for Switch-Level Verification of MOS Circuits," *IEEE Transactions on Computer-Aided Design*, vol. CAD-6, no. 5, pp. 766-779, September 1987.

[71] I. N. Hajj and D. G. Saab, "Fault Simulation and Logic Simulation of MOS VLSI Circuits Based on Logic Extraction," *Proceedings of the IEEE International Conference on Computer-Aided Design*, Santa Clara, California, pp. 99-100, November 1983.

[72] R. E. Bryant and M. D. Schuster, "Fault Simulation of MOS Digital Circuits," *VLSI Design*, pp. 24-30, 1983

[73] M. Kawai and J. P. Hayes, "An Experimental MOS Fault Simulation Program CSASIM," *Proceedings of the Design Automation Conference*, pp. 2-9, June 1984.

[74] P. Banerjee and J. A. Abraham, "A Multilevel Algebra for Modeling Physical Failures in MOS VLSI Circuits," *IEEE Transactions on Computer-Aided Design*, pp. 312-321, July 1985.

[75] H. H. Chen, R. G. Mathews, and J. A. Newkirk, "An Algorithm to Generate Tests for MOS Circuits at the Switch-Level," *Proceedings of the International Test Conference*, pp. 304-312, November 1985.

[76] I. N. Hajj and F. Najm, "Test Generation for Physical Faults in MOS VLSI Circuits," *COMP-EURO '87: VLSI and Computers*, Hamburg, Germany, pp. 386-389, May 1987.

[77] I. N. Hajj and D. G. Saab, "Switch-Level Logic Simulation of Digital Bipolar Circuits," *IEEE Transactions on Computer-Aided Design*, vol. CAD-6, no. 2, pp. 251-258, March 1987.

[78] V. B. Rao and T. N. Trick, "Network Partitioning and Ordering for MOS VLSI Circuits," *IEEE Transactions on Computer-Aided Design*, vol. CAD-6, no. 1, pp. 128-144, January 1987.

INDEX

ABOUT THE AUTHORS

Vasant B. Rao received his Ph. D. degree in Electrical Engineering from the University of Illinois at Urbana-Champaign in 1985. Since then, he is an Assistant Professor in the Department of Electrical and Computer Engineering at the University of Illinois at Urbana-Champaign. During the summer of 1987, he worked with Texas Instruments, Inc., Dallas, TX, as a Consultant. His research interests include the areas of simulation of VLSI circuits, stochastic algorithms for combinatorial optimization, and graph theory. He has written over 20 papers in these areas.

David V. Overhauser is currently a graduate student in the Department of Electrical and Computer Engineering and a Research Assistant at the Coordinated Science Laboratory, University of Illinois at Urbana-Champaign, pursuing a Ph. D. degree. In 1985 he was awarded an AT&T Bell Laboratories Ph. D. Research Scholarship. In 1979-80, he worked with Purdue University Computing Center, and in 1980-83, he worked with Micro Data Base Systems, Inc., Lafayette, IN. During the summers of 1982, 1986, and 1987, he was with IBM Corp., Poughkeepsie, NY, AT&T Bell Labs., Murray Hill, NJ, and Texas Instruments, Inc., Dallas, TX, respectively. His research interests include simulation of integrated circuits, parallel computing, and computer graphics.

Timothy N. Trick received his Ph. D. degree in Electrical Engineering from Purdue University, West Lafayette, IN, in 1966. Since 1965 he has been with the University of Illinois at Urbana-Champaign, where he is currently Department Head and Professor of Electrical and Computer Engineering. In the summers of 1970 and 1971, he was an ASEE-NASA Summer Faculty Fellow at the Marshall Space Flight Center, Huntsville, AL. In 1973-74, he was a Visiting Associate Professor with the Department of Electrical Engineering and Computer Sciences, University of California, Berkeley. During 1984-1986, he was Director of the Coordinated Science Laboratory, University of Illinois, Urbana. He is a Fellow of IEEE. In 1976, he received the Guellemin-Cauer Award for best paper published in the IEEE Transactions on Circuits and Systems, and in 1984 he was awarded the IEEE Centennial Medal. His research interests include computational methods for circuit analysis and design, integrated circuits, and analog and digital signal processing. He has over 120 publications.

Ibrahim N. Hajj received his Ph. D. degree in Electrical Engineering from the University of California, Berkeley in 1970. From 1970 to 1978, he was with the Department of Electrical Engineering at the University of Waterloo, Waterloo, Canada. He spent two years, form 1973 to 1975, teaching at the Lebanese University in Beirut, Lebanon. In 1978, he joined the University of Illinois at Urbana-Champaign, where he is now a Professor of Electrical and Computer Engineering. He served as an Associate Editor of the IEEE Transactions on Circuits and Systems, and is a member of the Computer-Aided Network Design (CANDE) Committee. His research interests include computer-aided design of VLSI circuits, simulation, testing, statistical analysis, and parallel processing. He has published over 90 papers.

GPSR Compliance

The European Union's (EU) General Product Safety Regulation (GPSR) is a set of rules that requires consumer products to be safe and our obligations to ensure this.

If you have any concerns about our products, you can contact us on ProductSafety@springernature.com

In case Publisher is established outside the EU, the EU authorized representative is:

Springer Nature Customer Service Center GmbH
Europaplatz 3
69115 Heidelberg, Germany

Batch number: 10682323

Printed by Printforce, the Netherlands